12th International Conference on Environmental Science and Technology (ICEST 2021)

IOP Conference Series: Earth and Environmental Science
Volume 1006

Online
24 – 26 September 2021

ISBN: 978-1-7138-5448-7
ISSN: 1755-1307

Printed from e-media with permission by:

Curran Associates, Inc.
57 Morehouse Lane
Red Hook, NY 12571

Some format issues inherent in the e-media version may also appear in this print version.

Institute of Physics
Dirac House, Temple Back
Bristol BS1 6BE UK

Phone: 44 1 17 929 7481
Fax: 44 1 17 920 0979

techtracking@iop.org

Additional copies of this publication are available from:

Curran Associates, Inc.
57 Morehouse Lane
Red Hook, NY 12571 USA
Phone: 845-758-0400
Fax: 845-758-2633
Email: curran@proceedings.com
Web: www.proceedings.com

TABLE OF CONTENTS

IOP Conf. Series: Earth and Environmental Science 1006 (2022) 011001 doi:10.1088/1755-1315/1006/1/011001

PREFACE

The 2021 12th International Conference on Environmental Science and Technology (ICEST 2021) was held successfully as virtual conference during September 24-26, 2021 due to the situation of pandemic. ICEST 2021 is a conference to bring together scientists, researchers, scholars, students and industry professionals to discuss the challenges, opportunities, problems and future development in the fields of environmental science and technology.

The location of organizer is Beijing Normal University. There are around 50 participants attending the conference in the opening ceremony and plenary sessions. They are from UK, Morocco, Ecuador, Korea, Portugal, Canada, Japan, China, and Thailand and so on. In this conference, we were greatly honored to have three outstanding Keynote Speakers and three Invited Speakers from different countries. They addressed to the audience and shared the knowledge and rich experience in related fields. They are Prof. Yongping Li from Beijing Normal University, China; Prof. Suiliang Huang from Nankai University, China; Prof. Yanpeng Li from Chang'an University, China; Assoc. Prof. Fengmin Wu from Chongqing Geomatics and Remote Sensing Center, China; Dr. Ong Chong Yong from Universiti Sains Malaysia, Malaysia and Dr. John Liu from Ontario Ministry of the Environment, Conservation and Parks, Canada & York University, Canada. There are 6 sub-sessions during the conference. They are Technologies for Environmental Pollution Control I & II, Modeling of Environmental Systems I & II, Sustainable Energy System and Engineering Infrastructure I & II. Attendees, paper presenters, keynote speakers, scholars and students have benefited in many ways from this conference.

We would like to express our great appreciation to the organizers, members of the program committees and reviewers of ICEST 2021 for their hard works. Without their participation, it would be impossible to hold ICEST 2021 successfully and ensure high quality of papers of the conference proceedings. We also would like to express our gratitude to the unbelievable number of authors for contributing their research results to the conference.

Yongping Li
Beijing Normal University, China
Dec. 7, 2021

IOP Conf. Series: Earth and Environmental Science 1006 (2022) 011001 doi:10.1088/1755-1315/1006/1/011001

Committees

Conference Chairs

Gordon Huang, University of Regina, Canada
Yongping Li, Beijing Normal University, China

Technical Program Committee

James T. Anderson, West Virginia University, USA
Guangwei Huang, Sophia University, Tokyo, Japan
Lixiao Zhang, Beijing Normal University, China

Publication Chair

Zhitong Xing, Peking University, China

Technical Members

Dina M. R. Mateus, Instituto Politécnico de Tomar, Portugal
Lovorka Gotal Dmitrović, University North, Croatia
Qiyuan Gu, General Research Institute for Nonferrous Metals Engineering, China
Liang Fang, Jiangping Bio-Substrate, China
Adam Fenech, University of Prince Edward Island, Canada
Vasilis Fthenakis, Columbia University, USA
Kailas Deoram Ahire, K.R.T. Arts, B.H. Commerce and A.M. Science (KTHM) College, Nashik, India
Kevin Fong-Rey Liu, Ming Chi University of Technology, China
Jen-Jeng Chen,Tajen University, China
Tae-Hyoung Gim, Seoul National University, South Korea
Radu Godina, NOVA University Lisbon, Portugal
Yan Hao, Beijing Normal University, China
Ryusuke Hashimura, International Institute of Natural Disaster, Japan
Henrique J. O. Pinho, Instituto Politécnico de Tomar, Portugal
Wendy Huang, McMaster University, Canada
Yee-Chang Jin, University of Regina, Canada
Vuttichai Kaewkrajang, Pollution Control Department, Thailand
Monzur Imteaz, Swinburne University of Technology, Australia
Zoe Li, McMaster University, Canada
Chuxia Lin, University of Salford, UK
Jinliang Liu, Ontario Ministry of Environment and Climate Change, Canada
Minghua Liu, Fuzhou University, China
Reza Nekovei, Texas A&M University-Kingsville, USA
Kriansak Pirarai, Department of Groundwater Resource, Thailand
Makridis Sofoklis, University of Patras, Greece
Vladimir Strezov, Macquarie University, Australia
Lerdphan Sukyirun, Office of the National Water Resources, Thailand
Qian Tan, China Agricultural University, China

IOP Conf. Series: Earth and Environmental Science 1006 (2022) 011001 doi:10.1088/1755-1315/1006/1/011001

IOP Conf. Series: Earth and Environmental Science 1006 (2022) 011001 doi:10.1088/1755-1315/1006/1/011001

Statement of Peer Review

All papers published in this volume of Earth and Environmental Science (EES) have been peer reviewed through processes administered by the proceedings Editors. Reviews were conducted by expert referees to the professional and scientific standards expected of a proceedings journal published by IOP Publishing.

IOP Conf. Series: Earth and Environmental Science 1006 (2022) 011002 doi:10.1088/1755-1315/1006/1/011002

Peer Review Statement

All papers published in this volume have been reviewed through processes administered by the Editors. Reviews were conducted by expert referees to the professional and scientific standards expected of a proceedings journal published by IOP Publishing.

- **Type of peer review:** Double Anonymous
- **Conference submission management system:** http://confsys.iconf.org/submission/icest2021
- **Number of submissions received:** 29
- **Number of submissions sent for review:** 27
- **Number of submissions accepted:** 14
- **Acceptance Rate (Submissions Accepted / Submissions Received × 100):** 48.3
- **Average number of reviews per paper:** 2.98
- **Total number of reviewers involved:** 30
- **Contact person for queries:**
  Name: Yi Xie
  Email: yix@pku.edu.cn
  Affiliation: Peking University - Environment

High-accuracy Machine Learning Models to Estimate above Ground Biomass over Tropical Closed Evergreen Forest Areas from Satellite Data

Ktawut Tappayuthpijarn[1] and Bernd S Vindevogel[2]

1 Lead Data Scientist, True Digital Group Co., Ltd., Bangkok, Thailand
2 Chief Analytics Officer, True Digital Group Co., Ltd., Bangkok, Thailand
E-mail: ktawut.tap@truedigital.com, bernd.vindevogel@truedigital.com

Abstract. Quantifying the amount of biomass stored in forested areas has been traditionally done with manual field measurements, which is costly, time consuming and doesn't scale well over large areas. This paper investigates the possibility to estimate the amount of Above Ground Biomass (AGB) using machine learning models with publicly available satellite data, where a large-scale training dataset has been created from a detailed biomass mapping project in the Democratic Republic of Congo (DRC). Several model architectures including the current state-of-the-art tree-based models were tested along with deep neural network (DNN) ones. It was found that DNN models provide slight improvement in accuracy, whilst can potentially be used for further fine-tuning with smaller local dataset for usage elsewhere outside of the DRC.

1.Introduction

The introduction of climate policies that put prices on CO2 emission and the establishment of carbon trading exchanges have begun to incentivised companies around the world to reduce their emissions. Charoen Pokphand (CP) Group plc., ltd., for example, aims to offset parts of its carbon footprint with several reforestation projects, planting millions of trees over thousands of hectares of forest before 2030 [1, 2]. Conservation projects such as this need to have their AGB quantified to issue carbon credits and use them to offset emission. Traditionally, this is done with field measurements on carefully-laid sampling plots by documenting tree species, their width and height, approximating the AGB within sampling plots with allometric equations, then extrapolating to cover the entire area [3]. However, such method is labour intensive, costly and difficult to scale. Often the required budget for field measurements is beyond most companies' reach, discouraging smaller ones from such activities and achieving humanity's ultimate goal of averting the worst effects of climate change. It's clear better means to estimate AGB over forested lands accurately, cheaply and at scale is needed.

1.1.Related works

Using drones or aircrafts equipped with Light Detection And Ranging (LiDAR) devices and / or cameras to estimate properties of forest canopy have been proposed in several works, e.g., [4-8]. This typically involves extracting canopy-related parameters such as fraction of canopy cover, height and / or species from images and / or LiDAR data. A notable work using this approach is [8] where Shapiro A et al. conducted hundreds of LiDAR survey flights over the Democratic Republic of Congo (DRC). The authors derived an allometric equation between the forest canopy structure and field-measured AGB data, then used it to create a detailed AGB map of the entire country. Although drone survey provides

IOP Conf. Series: Earth and Environmental Science 1006 (2022) 012001 doi:10.1088/1755-1315/1006/1/012001

accurate biomass estimation, it hasn't completely solved the cost and scalability issues as it still requires expensive equipments and conducting survey flights over the target area.

Another approach to estimate forest AGB is to use remote sensing data from various satellites and frequency bands, often at no cost. The related works employing this technique often follow a similar pattern - gathering various satellite data, pairing them with field AGB measurements and developing models to estimate the latter. How useful and accurate these models would be is often determined by a) the chosen bands and resolution of the data used as input features, b) the amount of field AGB data points used as targets for model training, and c) the models used to capture relationships between the two. Some notable works using this approach are discussed as follows.

Mitchard E et al. [9] built regression models to estimate AGB over shrublands in Mozambique using only L-band radar data with field AGB measurements. The resulting models were shown to be accurate, achieving ~70% accuracy and able to detect small changes of the AGB over time. Laosuwan T et al. [10] used optical images from the Landsat-7 satellite, estimated the crown cover fraction, and built regression models between the crown cover and approximately 50 field AGB measurements. Häme T et al. [11] used both optical and L-band radar data with simple regression models to estimate AGB. However available satellite sources back then could provide only low resolution data by today's standards. Coupled with limited field AGB data points the developed models only achieved ~60% accuracy on unseen data. A more recent work by Li Y et al. [12] explored using both C-band radar and optical data from the Sentinel-1 and Landsat-8 satellites to estimated AGB over China's subtropical forest regions. The authors investigated various linear and decision tree-based model architectures, and found that the XGBoost model outperformed all others, achieving ~60% accuracy. Dang A et al. [13] used Sentinel-2's optical data and several derived vegetation indices with a Random Forest (RF) model to estimate the AGB over tropical forest areas in Vietnam, achieving 80.45% accuracy. Lastly, Debastiani A et al. [14] used C-band and optical data from Sentinel-1, 2 along with several vegetation indices to achieve as much as 83.94% accuracy with a Robust Regression model.

1.2. Our contributions

Previous works with satellite data were limited to using simple statistical models, e.g., linear regression and tree-based ones partly due to having limited training data, often with only a few hundreds data points of field AGB measurements or less. This has prevented the usage of bigger DNN-based models [15] that are potentially better at capturing non-linear complex relationships, as the amount of training data should preferably be ~10 times the number of trainable parameters in the model [16]. To overcome this limitation, this paper provides the following contributions.

- A large dataset suitable for training large, deep neural network-based models was created. This dataset paired recent, free and publicly available satellite data with the detailed AGB map based on the work from Shapiro A et al. [8] to create 45,000 data points. The chosen satellite features will also be available by the current and future missions for several years to come, ensuring continuity and usability of the developed models in a medium / long term.
- AGB prediction models based on the current state-of-the-art [12-14], as well as DNN-based ones were trained and compared, where the latter ones showed slightly better performance.
- Although these DNN models were trained on the AGB map of the DRC, they can serve as pre-trained models for fine-tuning with smaller field AGB measurements elsewhere with similar forest structure in future works, e.g., over similar closed evergreen forest in Thailand.

The paper is structured as follows. Preparation of satellite data, derived features and the target AGB are covered in section 2. Section 3 provides information on the model architectures and their results are presented in section 4. Finally section 5 summarises the work and future improvement areas.

2. Data sources and feature preparation

2.1. Features from Sentinel-1 satellites

The European Space Agency (ESA)'s Sentinel-1 satellites provided C-band Synthetic Aperture Radar (SAR) backscattering data over any region on the Earth for free with a revisit frequency of approximately 5 days and would continue to be available until at least 2030. The data were available at

3 processing levels, in which the Level-1 Ground Range Detected (GRD) product obtained between Oct 2014 to Apr 2015 were used [17]. The C-band radiation was emitted in the Vertical polarisation. Features representing the backscattering power in the Vertical and Horizontal polarisations are denoted VV and VH respectively. Additionally, the Gray-Level Co-Occurrence Matrix (GLCM) with correlation calculation for the VV feature, and with second angular momentum calculation for the VH feature were also computed as 2 additional derived features, as they were suggested to have some correlations with the AGB by [14]. They are denoted as VV_corr and VH_asm respectively.

2.2. Features from Sentinel 2 satellites

The ESA's Sentinel-2 satellites provided optical and near-infrared data over any region on the Earth for free. They were due for decommissioned in 2022, but would be replaced by Sentinel-10 with better sensors in the same frequency range. The data were provided in multiple processing levels, in which the Level-2A Bottom of Atmosphere data with 11 frequency bands ranging from blue to Short Wave Infrared (SWIR) were used. These features, denoted as B1, B2, B3, B4, B5, B6, B7, B8, B8A, B11 and B12 were obtained between Jan 2018 to Jan 2021 [17].

2.3. Derived features

- Normalised difference indices. These are the normalised difference between the strength of any 2 features relative to their sum as shown in equation (1). The choices of features to use as *f1* and *f2*, their commonly known names and the information they provide are listed in table 1.

$$NormalisedDifferenceIndex = \frac{f1-f2}{f1+f2} \quad (1)$$

Table 1. A list of normalised difference indices

Name	Information revealed	*f1*	*f2*
Normalised Difference Vegetation Index (NDVI)	The presence of plants and their health	B8 (Near Infrared)	B4 (Red)
Normalised Difference Water Index 1 (NDWI1)	The presence of water content in plant and soil, an indicator of the ecological state of forest	B8A (Red Edge4)	B11 (SWIR1)
Normalised Difference Water Index 2 (NDWI2)		B8A (Red Edge4)	B12 (SWIR2)
Normalised Difference Infrared Index (NDII)		B8 (Near Infrared)	B11 (SWIR1)
Normalised Difference Cross-Polarisation Index (NDXI)	The presence of tree canopy, as suggested to be correlated by Vreugdenhil et al. [18]	VV	VH

- Enhanced Vegetation Index (EVI) is designed to reduce noise from atmospheric interference and be sensitive to forest canopy structure [19]. It was computed from Sentinel-2 features as.

$$EVI = 2.5 * \frac{NIR-R}{1+NIR+6R-7.5B} \quad (2)$$

where *NIR* is the Near Infrared (B8) feature
R is the Red (B4) feature
B is the Blue (B2) feature

- Leaf Area Index (LAI) is a derived feature provided by NASA, computed using data from its Terra and Aqua satellites [20] which would be available until at least 2026. It represents green leaf area per unit ground area in broadleaf forest canopy.
- Copernicus Global Land Cover classification (GLC) is a derived feature by ESA using data from its Sentinel missions. The feature provides detailed land type classification, e.g., urban, agricultural, closed evergreen broadleaf forest, closed deciduous broadleaf forest, etc [21].
- Copernicus Global Land Cover tree-cover fraction (GLC_tree_fraction), a derived feature by ESA using the same raw data as the GLC. It represents the fraction of land covered by tree.

Apart from these features, interaction features between them, e.g., by multiplying 2 features together to promote non-linearity were also explored. However none of them provided any gain in Mutual Information (to be discussed later) with the target AGB, thus they were not included for further study.

2.4. Carbon map of the Democratic Republic of Congo

To train large models, detailed and large AGB data are needed so they can be paired with satellite features in the same areas over the same period of time to create a large training dataset. The DRC's AGB map by Shapiro A et al. [8] is a good data source for this purpose. The map was available everywhere in the DRC at 1-hectare resolution (100x100m). It was constructed from field and high-accuracy LiDAR measurements in 2015 where data from Sentinel-1 and Sentinel-2 over the same area were also becoming available. Note that models trained on this dataset would fit to the forest types commonly found in the DRC, e.g., closed evergreen broadleaf forest according to the GLC feature.

2.5. Dataset preparation

A dataset of images was created by randomly sample 45,000 locations from within the DRC, where each location is 100x100m in size. The AGB of those locations were obtained from the AGB map discussed earlier in section 2.4. Other base and derived features discussed in section 2.1-2.3 over the same area were extracted and stacked on top of the AGB value (Figure 1a), creating images each with 24 input features (channels) and one target value. The resolution of these images is at 10m, i.e., there are 100 pixels in each image where each pixel covers a 10x10m area. This dataset, referred to as the *Image* dataset, is suitable for a variation of the DNN model, known as Convolutional Neural Network (CNN), to be discussed later in section 3. Another dataset was created from the *Image* dataset by computing key descriptive statistics (min, max, mean, variance, skew) of each feature in each image and concatenating them into a long numerical vector (Figure 1b). The latter one, referred to as the *Descriptive* dataset is suitable for all other model types that require inputs to be numerical vectors.

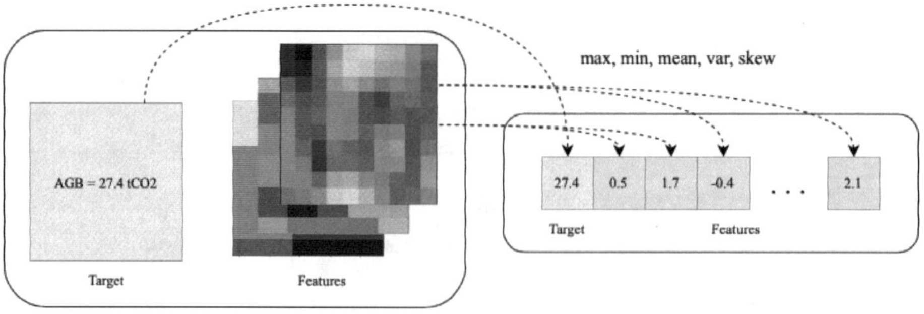

(a) A datapoint in the *Image dataset* (b) A datapoint in the *Descriptive dataset*

Figure 1. The *Image* dataset (a) and the *Descriptive* dataset (b) used for training models

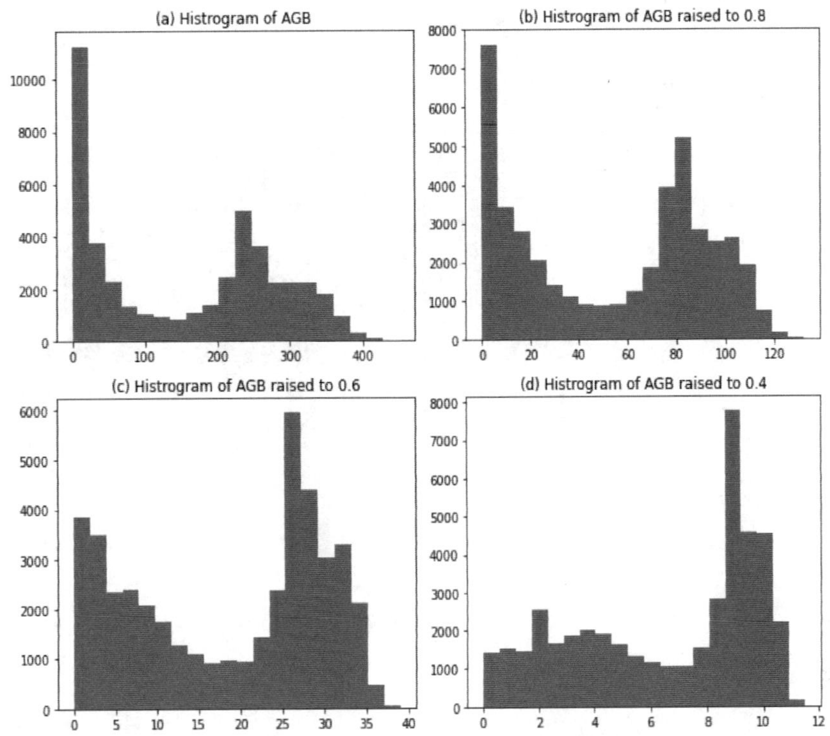

Figure 2. Histograms of the original and transformed AGB

The histogram of the AGB values is shown in Figure 2a where two peaks in the distribution are noticeable, one representing dense closed forests with higher AGB while another representing areas with lower AGB, e.g., open deciduous forest and grassland. By raising the target AGB values to the power of 0.8, 0.6 and 0.4 as depicted in Figure 2b, 2c and 2d respectively, the lower peak was spread wider while the higher peak was squeezed to be narrower. Models trained on these transformed AGB targets had to inverse transform their predictions - by raising them to the power of 1/0.8, 1/0.6 and 1/0.4 - to get the AGB value. However, the potential benefit was that it was easier for models to make more accurate prediction in the lower AGB range, at a cost of reduced accuracy on the higher end. Denote the *Image* and *Descriptive* datasets transformed by raising the power of the AGB value to 0.x as *Image_p0x* and *Descriptive_p0x* datasets respectively.

2.6. Mutual information with the target AGB data.

Mutual Information (MI) between 2 variables [22], also known as information gain, represents the amount of mutual dependency between them. It quantifies the amount of information obtained on one variable by observing the other with one advantage over simple correlation analysis in that it is able to capture non-linear dependencies between variables as well.

The MI values between the mean of features and the target AGB in the *Image* dataset were computed and shown in Figure 3. The best performing features providing large MI were mostly derived ones, e.g., the vegetation indices. Interestingly, Sentinel-1 features did not provide as much MI compared to others. This is in agreement with what was reported in other works [12, 14] that the C-band radar data provided little correlation with the AGB.

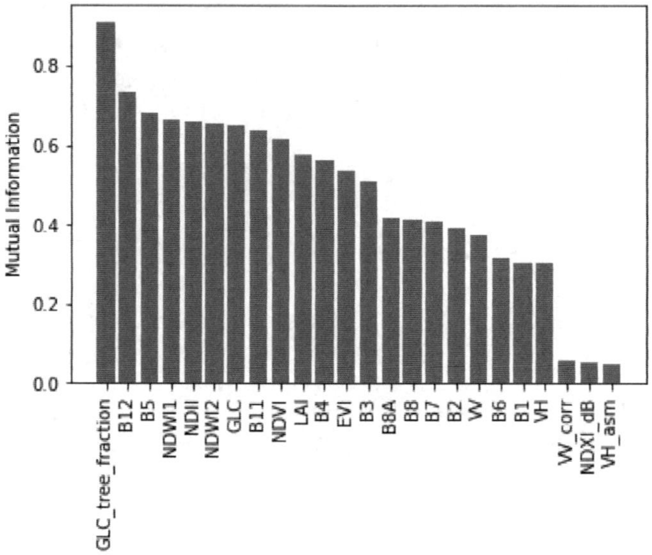

Figure 3. Mutual Information of each feature to the AGB target

3.Models

3.1.Linear regression model

The Linear Regression (LR) model assumes linear relationships between input features and the target variable. Specifically the target variable is approximated from an affine transformation of the input features. It is a simple model that requires a few data points to train, yet still performs quite well in many earlier works, e.g., by Debastiani A et al. [14] where a Robust Regression, a variation of the LR model achieved up to 83% accuracy.

3.2.Tree-based models

Tree-based models estimate the target variable by creating a set of decision criteria chained together or "tree", such that each node in the tree maximises the information gained with respect to the target variable successively until the leaf node is reached. There are several variations of the tree-based model, e.g., using various strategies to create the nodes, making an ensemble model from multiple weaker trees, etc. Earlier works, e.g., [12, 13] employed these models to achieve moderate AGB prediction accuracy. In this work, the XGBoost (XGB) [23] which is an ensemble model created from multiple weaker trees and trained using gradient boosting technique, and CatBoost (CB) [24] which is similar to XGB but with improved mechanisms to prevent overfitting were explored.

3.3.Deep neural network models

- Fully-Connected neural network models (FC) are DNN models with a simple structure, having layers of affine transformation followed by a non-linear transformation stacked on each other as shown in Figure 4a. It can be viewed as an extension to the logistic regression model by having multiple hidden layers in between, so the model can better capture more complex, non-linear relationships. A FC model with N layers is described by [L_1, L_1, …, L_N_t] where L_i is the size of the ith layer. L_N, the size of the last output layer is always 1.
- Convolutional Neural Network models (CNN) are DNN models designed to work with an image input. The following model description, also shown in Figure 4b, is a simple CNN architecture explored in this work. Given an input image with C_0 channels, a CNN model with the first layer having a kernel of size (k_1 x k_1 x C_1) applied the kernel over the input to create an output image with C_1 channels. The next layer with a kernel of size (k_2 x k_2 x C_2) further applied its kernel to

create the next image output with C_2 channels and so on. The size of these intermediate "images" could be reduced by using some sort of pooling strategy between layers, but it's not mandatory. Finally the output of these CNN layers was flatten and fed to fully-connected hidden layers of size L_i to predict a single-valued output as shown in Figure 3b. Since the images in the *Image* dataset were already quite small, having only 10 pixels per side, there was no need to use pooling layers. Thus, a CNN model in this work is fully described by a list $[(k_1, C_1), (k_2, C_2), \ldots, (k_c, C_c), L_{c+1}, L_{c+2}, \ldots, L_N]$ where L_N is always 1.

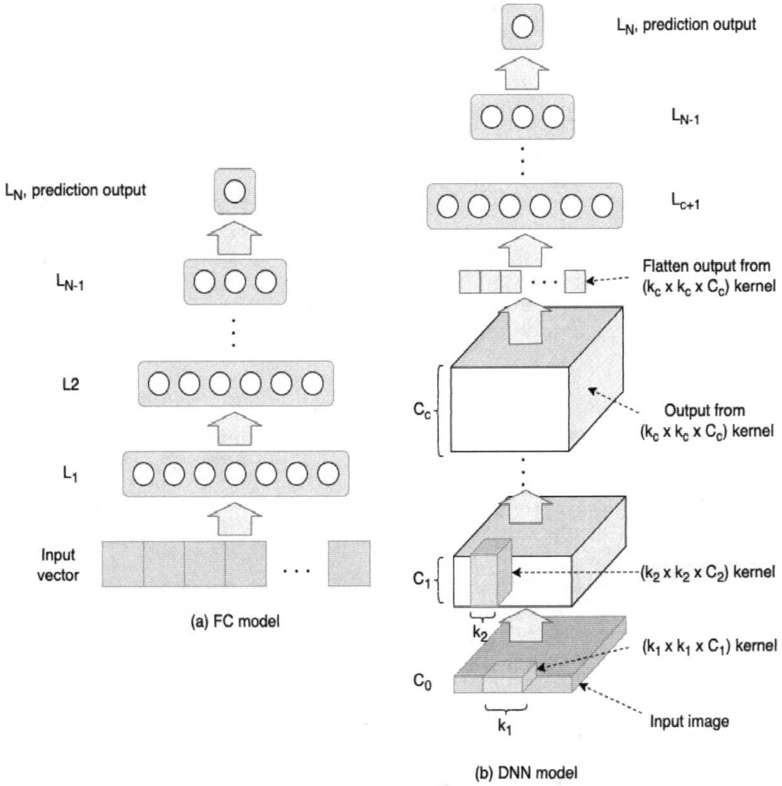

Figure 4. The explored DNN model architectures

4. Training results and discussions

The LR, XGB, CB and FC models were trained and their best hyper parameter sets were searched for on the *Descriptive_p04, Descriptive_p06, Descriptive_p08* and the *Descriptive* datasets. Similarly, the CNN model was trained and fine tuned on the *Image_p04, Image_p06, Image_p08* and the original *Image* datasets. These datasets were split at 9:1 ratio to create training and validation sets for hyper parameter tuning. The parameter search spaces for XGB and CB are based on what was recommended in [12], and also by stretching the libraries' default settings by some margins on both sides. The *min_mutual_information* filters only features with at least that much Mutual Information discussed in section 2.6 to be used. For FC and CNN models multiple layer combinations, kernel sizes, regularisation strengths and dropout rates were searched. The final fine-tuned models and their best parameter combinations are listed in table 2. These fine-tuned models were further evaluated on another unseen test area in the DRC, located between S2°37'44.03", E18°4'18.47" and S2°49'5.51", E18°20'37.31". This area had the GLC land classification as being 56.57% closed evergreen broadleaf forest with mean AGB, denoted as $AGB_{mean,\ clsd\_evr\_for}$ of 160.51 Mg/ha, while the mean AGB over the entire area with all land types, denoted as $AGB_{mean,\ all}$ is 99.67 Mg/ha.

Table 2. Parameter spaces and the best parameter set for all model types

Model architecture	Best parameters for all land types	Best parameters for closed evergreen broadleaf forest land type
Logistic Regression	winning_dataset=*Descriptive_p04* min_mutual_information=0.2	winning_dataset=*Descriptive_p04* min_mutual_information=0.2
XGBoost	winning_dataset=*Descriptive_p04* min_mutual_information=0.1 learning_rate=0.039 n_estimators=666 max_depth=9 min_child_weight=5 gamma=0.0 subsample=1.0 colsample_bytree=0.701 reg_alpha=0.5	winning_dataset=*Descriptive_p06* min_mutual_information=0.1 learning_rate=0.021 n_estimators=1000 max_depth=12 min_child_weight=5 gamma=0.061 subsample=0.791 colsample_bytree=0.772 reg_alpha=0.5
Catboost	winning_dataset=*Descriptive_p04* min_mutual_information=0.3 n_tree=1500 tree_depth=10 border_count=300 l2_leaf_reg=3.982	winning_dataset=*Descriptive_p08* min_mutual_information=0.15 n_tree=1296 tree_depth=9 border_count=232 l2_leaf_reg=3.720
FC	winning_dataset=*Descriptive_p08* min_mutual_information=0.35 layers=[150, 75, 30, 15, 7, 1] learning_rate=0.05 l2=0.05 dropout_p=0.05	winning_dataset=*Descriptive_p08* min_mutual_information=0.35 layers=[150, 75, 30, 15, 7, 1] learning_rate=0.05 l2=0.05 dropout_p=0.05
CNN	winning_dataset=*Descriptive_p06* min_mutual_information=0 layers=[(3, 60), 60, 1] learning_rate=0.01 l2=0.272 dropout_p=0.05	winning_dataset=*Descriptive_p08* min_mutual_information=0 layers=[(3, 60), 60, 1] learning_rate=0.01 l2=0.271 dropout_p=0.05

The AGB prediction accuracies with respect to $AGB_{mean,\ all}$ regardless of land type are shown in Figure 5a. When considering only predictions made over the closed evergreen broadleaf forest land type, the accuracies with respect to $AGB_{mean,\ clsd\_evr\_for}$ are shown in Figure 5b.

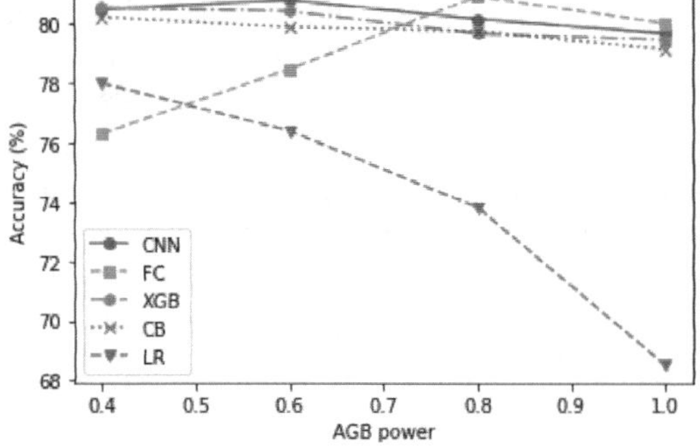

(a) Over all GLC land types

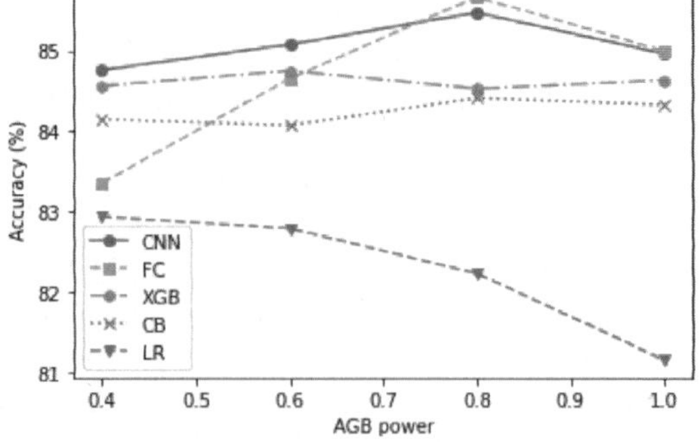

(b) Over closed evergreen broadleaf forest

Figure 5. AGB prediction accuracy

IOP Conf. Series: Earth and Environmental Science 1006 (2022) 012001 doi:10.1088/1755-1315/1006/1/012001

As can be seen in Figure 5, the accuracies of the FC and CNN models were better than all other models in both scenarios. Specifically when considering over all land types, the CNN model's accuracy was the highest at 80.812% (or equivalently at RMSE 19.125 Mg/ha) with the target AGB raised to the power of 0.6. For the best accuracy over closed evergreen broadleaf forest, the FC model had the best accuracy at 85.669% (or equivalently at RMSE 23.002 Mg/ha) with the AGB raised to the power of 0.8. No significant improvement was observed for tree-based models in both scenarios across the range of AGB transformation power. Note that the LR model also benefited from transforming the target AGB, even though it was still the weakest one in most cases. Finally the fine-tuned CNN model was used to create an AGB map over the test area in Figure 6b, and the ground truth AGB map of the same area is shown in Figure 6c for comparison.

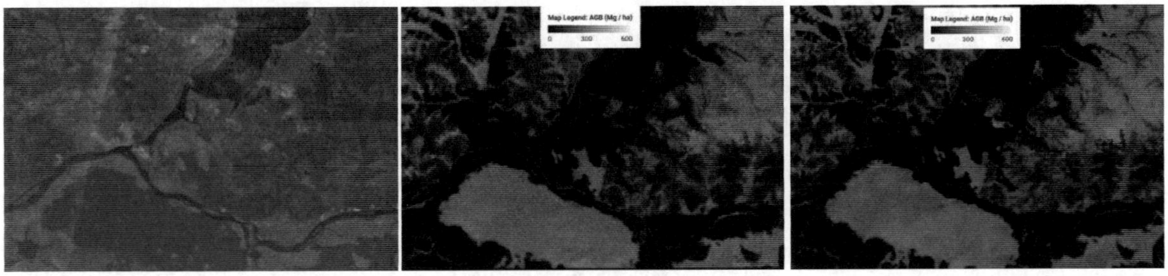

(a) RGB image (b) AGB map from CNN model (b) Ground-truth AGB

Figure 6. The test area

5.Summary and future works

In this work, a large dataset combining various free, currently available satellite features and the detailed AGB data over the DRC was created. The investigated features covered optical, Near Infrared, C-band SAR, derived vegetation and land classification indices. This allowed experimentation with various machine learning models including DNN for predicting the AGB.

It was found that the C-band and the derived features from it provided little Mutual Information with the AGB. Although similar observations were reported in earlier works [12, 14], it was also suggested longer-wavelength radar bands such as the L- and P-band that can penetrate forest canopy deeper and provide more structural information should be explored if they become available in the future.

To predict the AGB, two simple DNN-based models, along with a transformation trick to the target AGB were proposed. It was found that DNN models showed promising performances with accuracy ~80-85% over the closed evergreen broadleaf forest type, which is also the same dominant forest type found in other tropical countries. These DNN models can be used for further transfer learning to fit to smaller, but more localised and accurate AGB data from field measurements in future works.

6.References

[1] "CP Group joins global race for zero carbon emissions", *The Nation*, Oct 2020, https://www.nationthailand.com/news/30395459

[2] "Charoen Pokphand Group Drives Forward its Plan to Become a Carbon-Neutral Company by Planting Perennials throughout", *Bangkok Post*, Dec 2019, https://www.bangkokpost.com/thailand/pr/1820584/charoen-pokphand-group-drives-forward-its-plan-to-become-a-carbon-neutral-company-by-planting-perennials-throughout

[3] Estimation of carbon stocks and change in carbon stocks of trees and shrubs in A/R CDM project activities *UNFCC Clean Development Mechanism AR-Tool14*

[4] Lütjens B, Liebenwein L and Kramer K 2019 Machine learning-based estimation of forest carbon stocks to increase transparency of forest preservation efforts *arXiv* 1912.07850

[5] Lee J, Im J, Kim K and Quackenbush L 2018 Machine learning approaches for estimating forest stand height using plot-based observations and airborne LiDAR data *Forests* **9** 268

[6] Lim Y, La H, Park J, Lee M, Pyeon M and Kim J 2015 Calculation of tree height and canopy crown from drone images using segmentation *Journal of the Korean Society of Surveying, Geodesy, Photogrammetry and Cartography* **33** 605-613
[7] Kameyama S and Sugiura K 2020 Estimating tree height and volume using unmanned aerial vehicle photography and SfM technology, with verification of result accuracy *Drones* **4** 19
[8] Shapiro A, Saatchi S, Ferraz A, Xu L, Meyer V, Muanza E and Banks N 2017 High resolution carbon distribution in forests of Democratic Republic of Congo
[9] Mitchard E, Meir P, Ryan C, Woollen E, Williams M, Goodman L, Mucavele J, Watts P, Iain W and Saatchi S 2013 A novel application of satellite radar data: measuring carbon sequestration and detecting degradation in a community forestry project in Mozambique *Plant Ecology & Diversity* **6:1** 159-170
[10] Laosuwan T and Uttaruk Y 2016 Estimating above ground carbon capture using remote sensing technology in small scale agroforestry areas *Agriculture & Forestry* **62** 253-262
[11] Häme T, Rauste Y, Antropov O, Ahola H and Kilpi J 2013 Improved mapping of tropical forests with optical and SAR imagery, part II: Above Ground Biomass estimation *IEEE Journal of Selected Topics in Applied Earth Observations and Remote Sensing* **6** 92-101
[12] Li Y, L M, Li C and Liu Z 2020 Forest aboveground biomass estimation using Landsat 8 and Sentinel-1A data with machine learning algorithms *Sci Rep* **10** 9952
[13] Dang A, Nandy S, Srinet R, Luong N V, Ghosh S and Kumar S 2018 Forest aboveground biomass estimation using machine learning regression algorithm in Yok Don National Park, Vietnam *Ecological Informatics* **50** 1016
[14] Debastiani A, Sanquetta C, Corte A, Pinto N and Rex F 2019 Evaluating SAR-optical sensor fusion for aboveground biomass estimation in a Brazilian tropical forest *Annals of Forest Research* 10.15287/afr.2018.1267
[15] LeCun Y, Bengio Y and Hinton G 2015 Deep learning *Nature* 521(7553) 436-444
[16] Abu-Mostafa Y S 1995 Hints *Neural Comput* **7 (4)** 639–671
[17] Tappayuthpijarn K 2021 contains modified Copernicus Sentinel data 2014-2021, processed by ESA
[18] Vreugdenhil M, Navacchi C, Bauer-Marschallinger B, Hahn S, Steele-Dunne S, Pfeil I, Dorigo W and Wagner W 2020 Sentinel-1 Cross Ratio and Vegetation Optical Depth: A Comparison over Europe *Remote Sensing* **12**
[19] Huete A, Didan K, Miura T, Rodriguez E, Gao X and Ferreira L G 2002 Overview of the Radiometric and Biophysical Performance of the MODIS Vegetation Indices *Remote Sensing of Environment* **83** 195-213
[20] Myneni R, Knyazikhin Y and Park T 2015 MCD15A3H MODIS/Terra+Aqua Leaf Area Index/FPAR 4-day L4 Global 500m SIN Grid V006 [Data set_t]. NASA EOSDIS Land Processes DAAC. Accessed 2021-06-21 from https://doi.org/10.5067/MODIS/MCD15A3H.006
[21] Buchhorn M, Lesiv M, Tsendbazar N E, Herold M, Bertels L and Smets B 2020 Copernicus Global Land Cover Layers—Collection 2 *Remote Sensing* **12** 108, 1044
[22] Kreer J G 1957 A question of terminology *IRE Transactions on Information Theory* **3** 208
[23] Chen T and Guestrin C 2016 XGBoost: A Scalable Tree Boosting System *Proceedings of the 22nd ACM SIGKDD Intl Conf on Knowledge Discovery and Data Mining* 785–794
[24] Dorogush A V, Ershov V and Gulin A 2017 CatBoost: gradient boosting with categorical features support *Workshop on ML Systems at NIPS*

Design, Implementation and Evaluation of an Industrial Water Treatment System in El Coca, Ecuador in accordance with National Guidelines

Ricardo Urrutia-Goyes, Israel Moreano, William Bonilla and Stephany Mena

Departamento de Ciencias de la Energía y Mecánica, Universidad de las Fuerzas Armadas ESPE, Av. Gral. Rumiñahui s/n, P.O. Box 171-5-231B, Sangolqui 171103, Ecuador
E-mail: erurrutia@espe.edu.ec

Abstract. Water treatment systems are necessary in every industrial complex, especially when operations are held in the vicinity of a rainforest. The main focus of this study was to implement an industrial water treatment plant at the facilities of Halliburton company located in El Coca, Ecuador, in order to treat effluents and reuse them for machine washing and discharge to the environment. In first instance, evaluations were made of the operational conditions of an obsolete water treatment plant in the facilities, to analyze the requirements of a new treatment system. Physical characteristics were analyzed by identifying leaks, cracks and corrosion of the treatment units. Laboratory analyzes of the treated water were reviewed and parameters exceeding Ecuadorian guidelines (Ministerial Agreement N° 097-A) were identified. Once the requirement for a new treatment system was justified, evaluations of physical, chemical and organoleptic parameters of the wastewater were carried out, in which the existing pollutants were identified and treatment stages were established to remove the pollutants, complying with the beneficiary's technical and economic specifications. Once the industrial wastewater treatment plant was built, conditioning was carried out at each of its stages, as well as all the industrial mechanisms involved. Finally, monitoring of parameters was carried out every week for a period of one month to guarantee the correct performance of the treatment plant and compliance with the guidelines of the country.

1. Introduction

The development of human beings and technology has been a reason for environmental contamination through history. Both marine and freshwater environments have been contaminated and the extinction of beings that develop in these environments has been inevitable. Populations have also seen the proliferation of diseases that can be caused by the consumption of contaminated water [1,2]. Currently, there can be a spread of diseases due to the use of contaminated water, water shortages, or the lack of drinking water. Similarly, one of the biggest concerns for global governments is the quality of water since it can be contaminated by garbage residues, or industrial and municipal effluents, which are not treated properly. These effluents and well as other solid wastes are discharged to water bodies such as lakes, waterfalls and rivers. These waters once polluted, continue their journey causing an imbalance in aquatic environments especially to the animals that live in it and on its banks [3]. In current times, drinking water can contain potential pollutants that can be organic chemicals, inorganic chemicals, heavy metals, etc., from industrial, agricultural and urban runoff sources [2,3].
The oil industry is one that could cause the most negative impact on the environment and biodiversity. The operational practices that involve said industry include exploration, drilling, refining, transportation,

and consumption. All of these produce environmental deterioration due to the felling of trees, burning of fossil fuels, contamination of soil and water from oil spills, etc [4]. Companies in charge of each of these phases must be responsible and avoid or minimize said impacts through the implementation of technologies that reduce or eliminate the pollution produced.

Many of the environmental treatment or remediation techniques focus on the evaluation of physical, chemical and biological parameters from industrial wastewater. Such techniques seek the identification of pollutants with the highest concentrations and the proposition of solutions to reduce the environmental impacts of these discharges of these waters [4]. These methods are developed by carrying out evaluations that include obtaining data and sampling the residual effluent, where potential pollutants are subsequently identified using analytical monitoring equipment. For the design of a wastewater treatment system, aspects such as: flow, concentration of pollutants, available technologies, physical space, specialized personnel, and availability of economic resources must be taken into account [5].

In order to design a wastewater treatment train, physical methods such as the use of gravity, physical retention filtration, electrostatic attraction, etc. are commonly used as well as chemical processes such as oxidation, precipitation, coagulation, adsorption, etc. Biological methods involve the degradation or transformation of organic material by the action of microorganisms aerobically or anaerobically [2]. The implementation of any of these technologies is directly linked to the type of contamination and its respective concentration present in the water body that needs to be treated.

Industrial water treatment plants are capable of removing the vast majority of pollutants. These polluted water treatment systems are endowed with multiple physical, chemical and biological stages [6]. These systems have the capacity to regenerate effluents in a way that they do not have any negative effect on the environment anymore. Generally, the characteristics of these discharges are regulated by environmental organisms in every country. In Ecuador, Ministerial Agreement N° 097-A [1], details the limits for the discharge of contaminated water to a body of fresh water and states that municipalities will have the power to define maximum permissible loads to receiving bodies, the same that will be approved and validated by the national authorities and will be consigned in the discharge permits. In addition, it is established that discharges of untreated liquid wastes into receiving bodies, water conduction channels to reservoirs, irrigation channels or rain drainage channels, from the washing and/or maintenance of aerial and land vehicles, is prohibited, as well as that of manual and aerial applicators, containers, packages and containers that contain or have contained agrochemicals or other toxic substances [1].

The main goal of this study was to implement an industrial water treatment plant at the facilities of Halliburton company located in El Coca, Ecuador, in order to treat effluents and reuse them for machine washing and discharge to the environment according local regulations.

2. Methodology

This project was implemented in the facilities of Halliburton located in El Coca, Ecuador. An industrial water treatment plant was designed and built to comply with the parameters established by Ecuadorian guidelines, specifically Ministerial Agreement N° 097-A [1]. This allowed the reusage of treated water and its eventual release to the environment without negative impacts to the environment or surrounding areas.

2.1. Evaluation of physical parameters

The volumes of raw water flow that the industry produces daily were identified first. From this data all the treatment stages were dimensioned. The evaluations were carried out at the meeting point of all the effluents, which has a storage capacity of 2.5 cubic meters. The calculations were carried out over five days, taking a daily measurement at the beginning of working activities (8 AM) and at the end of the day (5 PM). The height values of the liquid were recorded and multiplied by the length and width dimensions, thus obtaining the volumetric value. To obtain the daily volume produced, the volume values measured at 5 PM were subtracted from those measured at 8 AM. These values are reported in Table 1.

For the determination of sedimentable solids, fats and oils, daily tests were carried out for five days at the end of the working day (5 PM). The tests were carried out y using a 1,000 ml graduated cylinder. Initially water samples from the meeting point of all the effluents were homogenized and 1,000 ml samples were taken in a graduated cylinder, they were left to rest for 30 minutes and the results were reported in terms of milliliters of the sedimentable solids content, as well as fats and oils, according to Ecuadorian technical standard NTE INEN 2169. These values can be seen in Table 2.

2.2. Evaluation of chemical parameters

The quantification of chemical parameters was carried out using HACH monitoring equipment with ISO 9001 certification. Temperature and pH measurements were carried out daily during the same 5-day period at the end of the working day (5 PM). In the analysis, HACH digital pH probes were used with integrated temperature sensors with the following characteristics: Accuracy ± 0.02 pH and range of 0-14 pH. Results are presented in Table 3. The evaluation of the BOD and COD parameters were carried out in the same daily fashion. HACH brand digital reactors were used under programming for standard digestions (BOD and COD) in which 20 ml of raw water (free of substances not soluble in water and sedimentable solids) were placed. The BOD and COD analyzes were performed separately. The process began by taking two 20 ml vials and filling them with raw water. Then the vials are placed in two different reactors and the option of COD analysis is selected for one reactor and BOD analysis for the other. The analysis is fully automatic and lasts between two to three hours. Once the analysis was finished, the reactor reported the values obtained on a digital panel. The BOD/COD ratio provided information about the biodegradability of the water and the methods for its treatment. These could be biological, chemical or a combination of both. Method 8000 ULR, LR, HR (HACH, 2016) [7]. The results are reported in Table 3.

Table 1. Calculation of daily industrial effluent flow.

Day	Liquid height 8 AM (m)	Volume of liquid 8 AM (m)	Liquid height 5 PM (m)	Volume of liquid 5 PM (m3)	Daily volume (m3)
1	0,400	0,625	1,870	2,922	2,297
2	0,350	0,547	1,500	2,344	1,797
3	0,520	0,813	2,200	3,438	2,625
4	0,250	0,391	1,250	1,953	1,563
5	0,380	0.594	1,690	2,641	2,047

Table 2. Quantification of sedimentable solids, fats and oils (sample size = 1 liter).

Día	Volume of settleable solids (ml)	Volume of fats and oils (ml)
1	50	10
2	20	8
3	100	5
4	75	16
5	45	10

Table 3. Quantification of temperature, pH, BOD and COD. Taken within 5 days.

Day	pH	Temperature (°C)	BOD (mg/L)	COD (mg/L)	BOD/COD
1	8,45	26,2	195	4,529	0.043
2	7,64	27,7	189	923	0.205
3	8,02	27,4	163	1,826	0.089
4	7,46	25,3	179	1,269	0.141
5	7,92	27,6	155	1,496	0.104

For the quantification of dissolved solids, total nitrogen, phosphates, surfactants, HACH brand DR 900 multiparameter meters with digital probes were used.
Dissolved solids and electrical conductivity analysis were carried out by placing 100 ml of raw water free of non-water soluble substances and settleable solids in a beaker. Coupled digital probes were placed on the multiparameter meter and analyzes were carried out by choosing the parameter to be analyzed on the digital panel. These measurements were made daily during the 5-day period at the end of the working day (5 PM). Method 8166 (HACH, 2016) [7]. The data are reported in Table 4.
For total nitrogen, phosphates and surfactants. The analyzes were carried out by placing 10 ml of raw water in the test vials. The 10 ml of raw water were previously treated by separating funnels and filtration with filter paper, so that the sample is free of sedimentable solids and substances not soluble in water. The vials were placed in the multiparameter meter and the analyzes were carried out by choosing the parameter to be analyzed on the digital panel. These measurements were made daily during the 5-day period at the end of the working day (5 PM). Method 8192, 8007, 8028 was followed [7]. These data are reported in Table 4.

2.3. Design of the industrial water treatment train

The treatment train constitutes a set of physical, chemical and biological processes, whose objective is to gradually remove the contaminants previously evaluated [8]. The designed water treatment train can be seen in Fig.1 and the hydraulic design, pipes, pumps, and valves can be seen in Fig 2. Both figures were developed according to the following criteria:
Through the evaluation of physical parameters reported in Table 2 related to the content of settleable solids and non-soluble pollutants in water (fats, oils and oil), the industrial wastewater sanitation process includes a pretreatment which aims to eliminate said pollutants. Due to the characteristics of the flow, the ideal pretreatment for this case was a grease trap. This stage provides an effluent free of settleable solids and non-water soluble agents.
An implanted primary treatment removes colloidal and dissolved solids, heavy metals, strong acids, and bases. For the removal of said contaminants, coagulation and flocculation units (chemical treatment) were chosen. The coagulation stage was responsible for neutralizing ions present in the treatment water (colloidal solids, dissolved, salts and heavy metals) causing them to agglutinate (as a reaction with aluminum polychloride). Flocculation was produced by supplying the flow with a chemical flocculating agent (pH - HACH). This allows the particles to agglutinate more quickly, facilitating sedimentation.
The next step was the removal of the organic load present in the treatment water. In this regard, the BOD values obtained in the evaluation of the chemical parameters were analyzed (which indicated that they were slightly out of the norm). In order to reduce said organic load, stages that involve aerobic reactors were implemented. This process consists of supplying dissolved oxygen to the treatment water through the action of compressors and bubble diffusers. The oxygen allows the activation of aerobic bacteria, which consume the organic load present. This system has removal efficiencies of up to 96% and only produces active sludge, which can be used later as fertilizer.

Table 4. Evaluation of conductivity, total nitrogen, phosphates and surfactants. Taken within 5 days.

Day	Conductivity (uS/cm)	Total nitrogen (mg/L)	Total phosphates (mg/L)	Surfactants (mg/L)
1	680	15	5.17	2,062
2	522	10	6.67	1,875
3	759	12	5.38	2,037
4	510	10	3.74	2,397
5	580	13	5.52	1,823

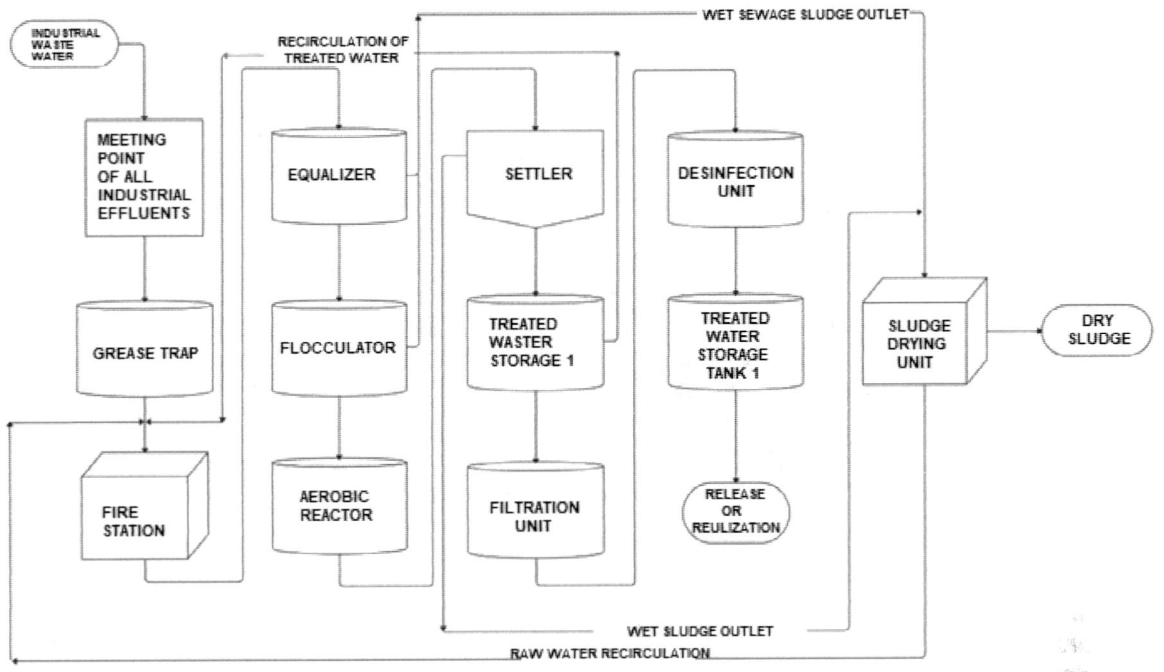

Figure 1. Industrial water treatment train.

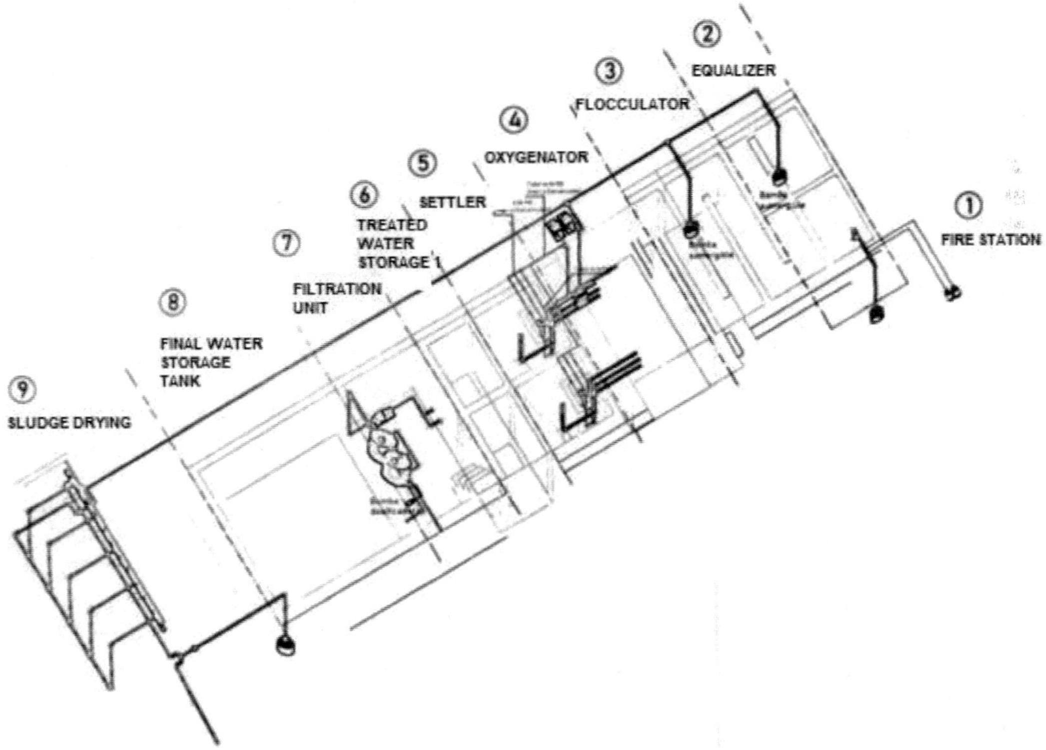

Figure 2. Hydraulic design, pipes, pumps and valves.

IOP Conf. Series: Earth and Environmental Science 1006 (2022) 012002 doi:10.1088/1755-1315/1006/1/012002

Effluent water from the aerobic reactor passes to the sedimentation stage, where one part of the active sludge produced in the previous stage is recirculated and another part is settled. The sedimentation technique consists of a conical pool equipped with a system of pipes that allow the recirculation of the flow. This stage operates through the action of compressors and control valves.

The sludge produced in coagulation, flocculation, and aerobic reactor stages are evacuated to the sludge drying chamber, using submersible industrial pumps. These chambers are equipped with filtering layers of different grain size, which prevent the passage of solids and allow the water contained in the sludge to permeate. The effluent water from this stage is recirculated to the coagulation stage.

Water coming from the settler must have properties such as: odorless, colorless, neutral pH, low levels of COD, BOD and low level of soluble solids. From this point on in the process, the water is called "treated water" and passes to a storage stage in which it can be sent to the pumping station if the wastewater presents excessive levels of BOD, COD, soluble solids, acidity. and basicity. I could also be sent to the filtration unit, where it is subjected to activated carbon and zeolite sand filters in order to remove heavy metals, surfactants and bacteria.

The treated water is then subjected to a disinfection process in which sodium hypochlorite is used, by means of dosing pumps. This method is used because it is inexpensive, effective, and easy to operate, unlike other methods, which are effective but of high economic demand, such as ozonation or the use of ultraviolet light.

Finally, the treatment process ends in a final water storage tank in which it can be reused for washing well drilling equipment, cars, trucks, mobile diesel production systems, and can also be released into the environment without causing any kind of environmental impact.

3. Results and discussion

The start-up of the plant took place during a period of one month with working hours from 8 AM to 5 PM from Monday to Sunday. All labor measures in the company were complied. The objective of this period was to adjust operating parameters and train the technical personnel who will be in charge of operating the plant. Four analyzes were carried out weekly during this period. The parameters evaluated are those described by Ecuadorian Ministerial Agreement N° 097-A [1, 9, 10]. Monitoring the treated water during a period of one month lead to prove that the industrial water treatment plant provides concentrations lower than those established in the mentioned Ecuadorian guideline. Obtained parameters can be seen in Table 5.

Throughout the month of operation, all the parameters analyzed reported values below the applied standard. There was only one exception: surfactants analyzed in week 3 showed a value that exceeded the limit by 0.11 mg/L. This occurred due to an unexpected increase in machine washing, and for such activitiy, additional industrial detergents were used to remove all the grease and dirt present. Therefore, the use biodegradable detergents are recommended to be used.

4. Conclusions

In the present work, an industrial water treatment system from Halliburton located at El Coca, Ecuador was designed, implemented and evaluated, according to Ecuadorian guidelines. The main point in designing the industrial water treatment plant was the physical, chemical, and biological characterization of the water to be treated, since the most suitable sanitation methods to treat the wastewater were selected from these data. The physical, chemical, and organoleptic parameters of the water to be treated were analyzed. With this, the polluting agents with the greatest presence were found, such as: chemical oxygen demand, surfactants, electrical conductivity, dissolved solids, total hydrocarbons, and phenols. This led to the elaboration of a flow chart of the process which has pretreatment, primary, secondary and tertiary treatments that are part of the whole treatment train. What most influenced the chemical analysis is the correct preparation of the raw water samples, since the presence of oils, grease, petroleum, and sedimentable solids cause interference in the monitoring equipment, which causes reports of erroneous results.

The industrial water treatment plant was implemented and it was evaluated for compliance of the environmental quality and effluent discharge regulations In Ecuador [1]. The most important part of the implementation is the development of the process flow chart since it must include treatment stages capable of removing the pollutants present in the analyzed wastewater,. The process flow chart developed is specific to treat the industrial waters of the company, because the presence of different pollutants and their respective concentration vary according to the work carried out by each company. The treated water was evaluated according to the regulations of the environmental management law for the prevention and control of environmental pollution. All the evaluated parameters presented values lower than the established discharge limits. The most important thing in the evaluation of the treated water was the correct operation of the industrial water treatment plant because an error in the dosage of chemical agents, an incorrect handling of the backflows, or an excess of pollutants can cause failures in various stages of the process treatment, which would cause the evaluated water not to meet the quality standards governed by the norm. What most helped in the correct management of the plant is continuous monitoring, since a deviation from previously established parameters can be remedied with action.

Table 5. Data obtained from the analysis of treated water during the period of one month of evaluation.

Parameters	Expressed as	Unit	Maximum allowable limit	Week 1	Week 2	Week 3	Week 4
Barium	Ba	mg/L	2.0	0.30	0.18	0.42	0.27
Cadmium	Cd	mg/L	0.02	0.01	0.01	0.01	0.01
Ammonia Nitrogen	N	mg/L	30.0	1.12	5.89	2.37	2.81
Chlorides	Cl	mg/L	1,000	381.41	420.18	453.23	356.12
Copper	Cu	mg/L	1.0	0.20	0.12	0.46	0.28
Total chromium	Cr	mg/L	0.5	0.10	0.21	0.12	0.10
Biochemical Oxygen Demand	BOD5	mg/L	100	30.00	23.00	18.00	12.00
Phenols	Fenol	mg/L	0.2	0.05	0.05	0.1	0.07
Total mercury	Hg	mg/L	0.005	0.002	0.002	0.002	0.002
Nickel	Ni	mg/L	2.0	0.10	0.09	0.09	0.16
Dissolved oxygen	O	mg/L	**	6.20	8.12	5.12	7.89
Chemical oxygen demand	COD	mg/L	200	120	98	180	112
Vanadium	V	mg/L	**	0.40	0.29	0.47	0.32
Lead	Pb	mg/L	0.2	0.15	0.12	0.16	0.10
Hydrogen potential	pH		6-9	6.83	6.59	6.30	7.2
Total solids	TS	mg/L	1,600	910.51	945.16	1,120.83	830.97
Sulfates	SO4	mg/L	1,000	46.00	57.19	112.03	52.18
Total hydrocarbons	TPH	mg/L	20.0	1.20	1.18	2.30	2.21
Surfactants	Active Substances to methylene blue	mg/L	0.5	0.31	0.15	0.61	0.22
Zinc	Zn	mg/L	5.0	0.05	0.05	0.08	0.06

IOP Conf. Series: Earth and Environmental Science 1006 (2022) 012002 doi:10.1088/1755-1315/1006/1/012002

5. References

[1] M. J. Cabrera, A. Jácome, M. J. Carrillo, L. Campaña, D. Barragán, M. Enríquez, R. Rodríguez, P. Carrera, K. Vásconez, R. Guaña y L. Tapia, "Acuerdo Ministerial No. 097-A," Ministerio del Ambiente, Quito, 2015.

[2] M. Vega Loyola, "Diseño de plantas de tratamiento de aguas residuales," Tesoem, México, 2010.

[3] S. E. Manahan, "Introducción a la Química Ambiental", Mexico, D.F: Reverté Ediciones, S.A. de C.V y Editorial Reverté,S.A., 2007.

[4] F. Osorio, J. C. Torres y M. Sánchez, "Tratamiento de aguas para la eliminación de microorganismos y agentes contaminantes," Madrid: Ediciones Díaz de Santos, 2010.

[5] A. Noyola, J. M. Morgan y L. P. Guereca, "Selección de tecnologías para el tratamiento de aguas residuales municipales", Mexico: Instituto de ingeniería UNAM, 2013.

[6] J. H. Arocutipa, "Evaluación y propuesta técnica de una planta de tratamiento de aguas residuales en Massiapo del distrito de alto Inambari - Samdia," Universidad Nacional del Altiplano, Puno, 2013.

[7] HACH, "DR900 Colorimeter," 20 Abril 2016.

[8] A. R. Fernández, R. García Rosal y M. Dorado Valiño, "Tratamiento avanzados de aguas residuales industriales", Madrid: Dirección General de Universidades e Investigación , 2006.

[9] Acuerdo ministerial N° 097-A, "Registro oficial," Noviembre 2015.

[10] Peisol, "Ingeniería y soluciones de proyectos integrales," Peisol, Lago Agrio, 2017.

Acknowledgments

The current research efforts are part of ESPE Research Projects (No. 2020-PIC-012-CTE).

Climate Change Impacts on Agriculture and Water Resource Management in the North China Plain

Yanwei Lin[1] and Guangyu Li[2]

1 Department of Life Sciences, Imperial College London, UK
2 Institute for Atmospheric and Climate Science, ETH Zurich, Switzerland
E-mail: y.lin20@imperial.ac.uk

Abstract. North China Plain (NCP) is a main agricultural production region in China, dominated by a summer maize-winter wheat double cropping system. However, the grain yield highly depends on groundwater extraction to fulfil the irrigation demand. Additionally, the warming and drying tendency under climate change will further challenge water resource availability and agricultural productivity. This study applied modelled regional precipitation and temperature within the East Asia domain from the Coordinated Regional Climate Downscaling Experiment to derive climate change indicators and quantify future water resource utilisation. Results show a 30% increase in water irrigation demand (IWD) for growing maize in Jiangsu Province and a slight decrease in central Shandong Province. While for wheat, IWD decreases by 3-24% over the NCP except for northern Hebei Province and northern Shandong Province. Therefore, we suggest reducing agricultural fields for summer maize in southern NCP and winter wheat in northern NCP, and intensifying cultivation in central Shandong Province. Mann-Kendall statistics indicate an indistinct trend for IWD but a significant increase in crop evapotranspiration (ET_c) during 2021-2050, highlighting the necessity to cope with rising ET_c and apply varieties with high stress-tolerance to release irrigating water intensity under future climate change.

1. Introduction

North China Plain (NCP) is the largest and most important agricultural region in China [1], supporting 75% of total wheat production and 35% of maize production [2]. The leading cropping system is wheat-maize rotation, where annual water consumption for winter wheat is 450 mm and for summer maize is 360 mm [3]. However, over 70% of the total rainfall occurs in maize growing seasons, leaving a 200-300 mm water deficit for wheat seedlings and maturing [3]. Thus, the sustainability of grain yield highly depends on irrigation, and groundwater resources are intensively exploited to compensate for surface water shortage, resulting in a detrimental water-table decline and water quality degradation [2, 3]. Over the past 50 years, the water crisis was exacerbated by climate change, where annual precipitation has decreased by 6.7% [4], and increasing evapotranspiration and extreme droughts occurred [2]. To plan for sustainable water use, it is vital to predict future climate change and assess its impacts on regional hydrological system and agricultural activities.

General circulation models (GCMs) are valuable tools to project large-scale and long-term climate under different emission scenarios [5]. However, the coarse spatial resolution and inability to simulate subgrid topographical forcing make them insensitive to local processes, particularly precipitation [5, 6]. Regional climate models (RCMs) are thus developed to address important terrain effects and small-scale climate patterns, which is more realistic and preferable to regional projections [7]. However, lacking studies integrated high-resolved RCMs that reduce model uncertainties and improve data reliability. This study applied modelled regional precipitation and temperature to derive climate change indicators including effective precipitation (P_e), crop evapotranspiration (ET_c) and irrigation water demand (IWD). Additionally, we discussed the spatial distribution of IWD change and adaptation strategies for agriculture in the NCP.

2. Materials and methods

2.1. RCMs and historical observations

Modelled regional precipitation and near-surface air temperature were obtained from three RCMs participating in the Coordinated Regional Climate Downscaling Experiment (CORDEX) [8]: Hadley Centre Global Environmental Model-Regional Atmospheric Model (HadGEM3-RA), Regional Climate Model version 4 (RegCM4), and Yonsei University - Regional Spectral Model (YSU-RSM). Raw model outputs were calibrated based on historical observations of precipitation and temperature from the Asian Precipitation - Highly-Resolved Observational Data Integration Towards Evaluation [9] and National Centre for Atmospheric Research [10], respectively.

2.2. Bias correction

Temperature was corrected by the mean-based method [Equation (1)], assuming equal variance between observations and model simulations. Precipitation was calibrated by the variance-based method [Equation (2)], fitting both the mean and standard deviation of observations.

$$T_{BC} = T_{RAW} + (\overline{O_{REF}} - \overline{T_{REF}}) \tag{1}$$

$$P_{BC} = \overline{O_{REF}} + \frac{\sigma_{O,REF}}{\sigma_{P,REF}} (P_{RAW} - \overline{P_{REF}}) \tag{2}$$

where T and P represent temperature and precipitation. O_{REF} and T_{REF}/P_{REF} are observations and model outputs in reference periods, respectively. T_{RAW}/P_{RAW} and T_{BC}/P_{BC} represent raw model outputs and bias-corrected outputs in prediction periods, respectively. σ is the standard deviation and the bars above symbols are the mean.

2.3. Climate change indicators

P_e, ET_c and IWD were derived as climate change indicators to quantify regional water budget and water resource utilisation. P_e is the fraction of precipitation available for crops, excluding surface runoff, deep percolation and evaporation [11]. It was calculated by Equation (3).

$$P_e = \mu\, P \tag{3}$$

where μ is effective utilisation coefficients of precipitation in the NCP (Table 1). ET_c is the combination of plant transpiration and soil evaporation at different crop growth stages. It was calculated by Equation (4).

$$ET_c = K_c\, ET_0 \tag{4}$$

where K_c is crop coefficients in the NCP (Table 2) and ET_0 is potential evapotranspiration estimated by the Thornthwaite method [12]. IWD is the difference between crop water requirements and the effective precipitation [Equation (5)], where component of leaching requirements is neglected in the NCP due to low groundwater level and soil water storage [11].

$$IWD = \begin{cases} ET_c - P_e & ET_c > P_e \\ 0 & ET_c \le P_e \end{cases} \tag{5}$$

Table 1. Effective utilisation coefficients of precipitation in the NCP.

Monthly precipitation (mm)	μ
< 5	1.00
5-30	0.85
30-50	0.80
50-100	0.70
100-150	0.58
> 150	0.48

Table 2. Crop coefficients at different crop growth stages in the NCP.

	Growth stage	Month	Crop coefficient
Winter wheat	Initial season	October	0.69
	Mid-season	March, April	1.17
	End of late-season	May	0.34
Summer maize	Initial season	June	0.76
	Mid-season	July, August	1.13
	End of late-season	September	0.43

2.4. Time-series analysis

Nonparametric Mann-Kendall (MK) test [13, 14] was applied to quantify the significance in trends in data series at 95% confidence interval. Positive values of the test statistics z indicate increasing trends, whereas negative values indicate decreasing trends, with $|z| > 1.96$ indicating the statistical significance.

3. Results

3.1. Climate change impacts on water resources

Figure 1 shows an increase in IWD over the NCP compared to historical data. During wheat growing seasons, P_e is comparable with ET_c and IWD is relatively low (60.3 ± 20.2 mm/year). While for maize growing seasons, the amount of P_e cannot compensate ET_c, and IWD reaches 312.8 ± 47.8 mm/year. Figure 2 shows the trends of climate change indicators during 2021-2050. MK statistics indicate indistinct trends for P_e and IWD but a significant increase in ET_c in maize mid-season.

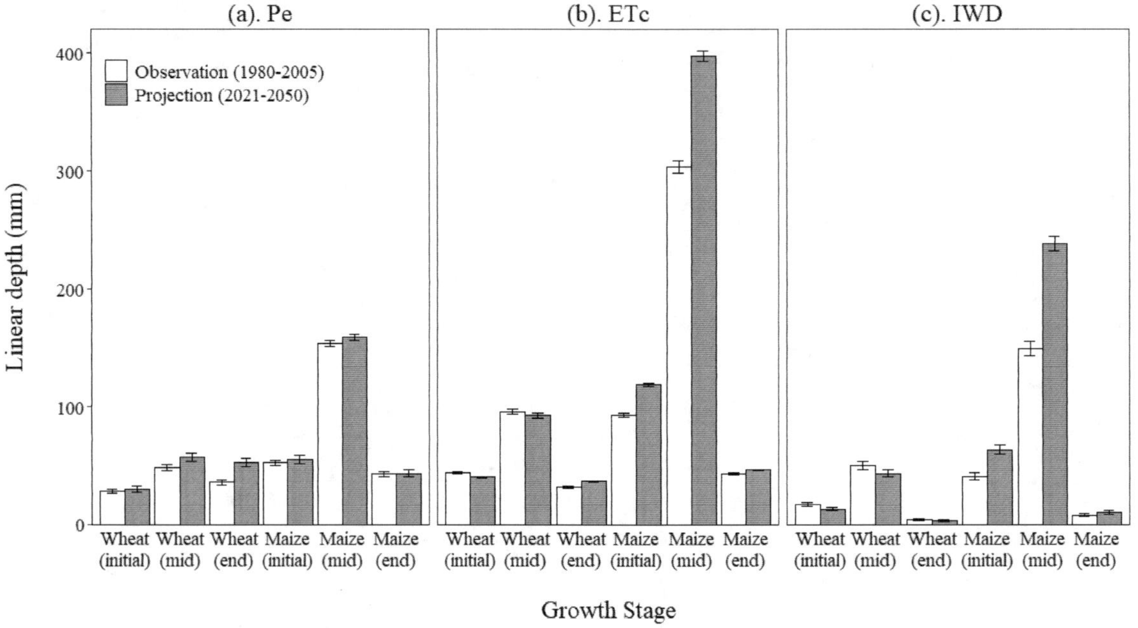

Figure 1. Climate change indicators in different crop growth stages. Data are derived from the mean of the three RCMs. Error bars are the standard error.

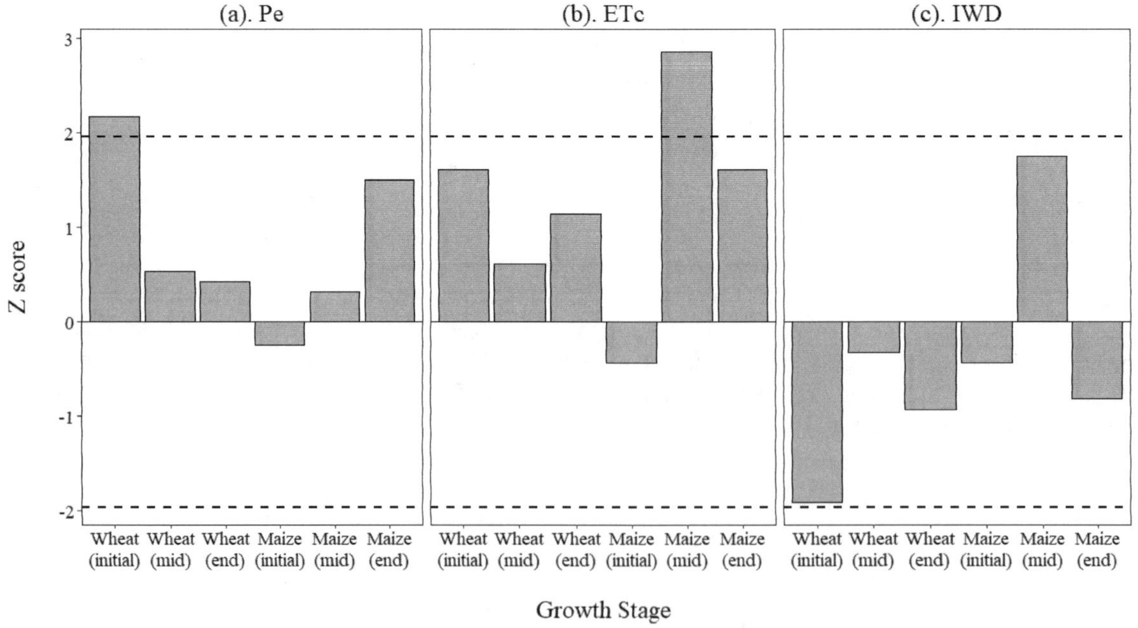

Figure 2. MK test statistics for climate change indicators during 2021-2050. Data are derived from the mean of the three RCMs. Dashed lines are the threshold of statistical significance of 95% confidence interval.

3.2. Spatial distribution of IWD

Figure 3 shows the spatial distribution of IWD change in different RCMs. An overall increasing trend in maize IWD is consistent among the three RCMs, with up to 30% increase in Jiangsu Province (Figure 3 (a), (c), (e)) and a slight decrease in central Shandong Province (Figure 3 (a), (c)). While for wheat IWD, RCMs project distinct trends, where 3-24% decrease is found in most parts of NCP in RegCM4 and YSU-RSM (Figure 3 (d), (f)) but HadGEM3-RA shows an increasing trend of 10-33% except for Jiangsu Province (Figure 3 (b)). Additionally, HadGEM3-RA and YSU-RSM highlight an approximately 40% increase in the north and northeast, respectively (Figure 3 (b), (f)).

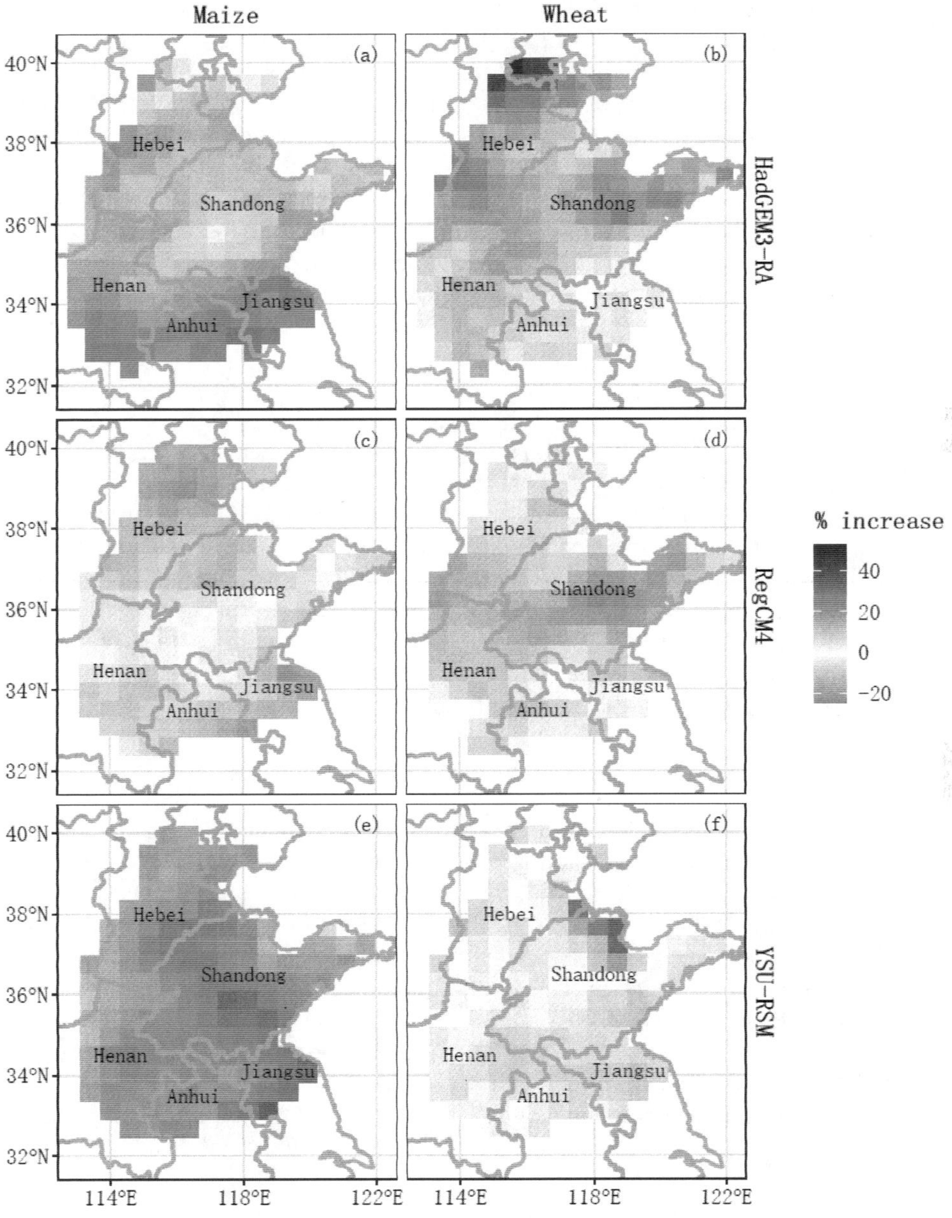

Figure 3. Spatial distribution of IWD change relative to historical data (1980-2005).

4. Discussions

Temperature is a dominant driver of evaporation and transpiration by transferring energy and affecting stomatal opening, while precipitation directly regulates soil moisture and crop water cycle [15]. Future warmer conditions in summer monsoon can increase soil evaporation and weaken monsoon circulation, reducing soil moisture and summer rainfall [16, 17]. This period is also the jointing stage to early grain filling stage for summer maize, where two-thirds of total evapotranspiration occurs [16]. Thus, low soil moisture and high evapotranspiration demand result in drought stress and substantial maize IWD. In contrast, the growth of winter precipitation can increase water availability for wheat and alleviate potential risks of evapotranspiration-derived water shortage [4].

Although indistinct trends in IWD are found for both crops during 2021-2050, a statistically significant increase in ET_c still highlights the necessity to cope with high ET_c and store water for additional evapotranspiration demand. Crop varieties with high thermal and drought tolerance can enhance water use efficiency and make full use of heat resources under climate change [2, 15]. They can also be sown more densely and reduce the total growing area to agricultural water consumption [18]. Therefore, in the region where water resources are insufficient, such as the NCP, stress-tolerant varieties and less cultivation area are feasible options to optimise water use and minimise yield loss under climate change.

Based on the south-north gradient in IWD change, we suggest reducing wheat production in northern Hebei Province and intensifying cultivation in central or southern Shandong Province to minimise total water consumption. Similarly, agricultural fields for summer maize need to decrease in Jiangsu and Anhui Province, and can potentially expand in central Shandong Province. Additionally, cropping system can be converted to specialise more in winter wheat as less IWD is expected, and select maize varieties with higher water uptake efficiency and drought tolerance.

5. Conclusions

In this study, we applied modelled regional precipitation and temperature from three RCMs participating in the CORDEX to derive climate change indicators, including P_e, ET_c and IWD. We quantified regional water budget and water resource utilisation in the NCP during 2021-2050. Wheat IWD is relatively low (60.3 ± 20.2 mm/year), while maize IWD shows a substantial amount (312.8 ± 47.8 mm/year). The spatial distribution of maize IWD indicates a 30% increase in Jiangsu Province and a slight decrease in central Shandong Province. Wheat IWD decreases by 3-24% over the NCP except for northern Hebei and northern Shandong. Therefore, we suggest a). applying crop varieties with high stress-tolerance, b). reducing agricultural fields for summer maize in southern NCP and winter wheat in northern NCP, and c). intensifying cultivation in central Shandong Province under future climate change.

6. References

[1] Li X, Ju H, Sarah G, Yan C, William D B and Liu Q 2017 Spatiotemporal variation of drought characteristics in the Huang-Huai-Hai Plain, China under the climate change scenario *Journal of Integrative Agriculture* **16(10)** 2308-2322

[2] Mo X, Hu S, Lin Z, Liu S and Xia J 2017 Impacts of climate change on agricultural water resources and adaptation on the NCP *Advances in Climate Change Research* **8** 93-98

[3] Liu C, Yu J and Eloise K 2001 Groundwater exploitation and its impact on the environment in the North China Plain *Water International* **26** 265-272

[4] Fu G, Charles S P, Yu J and Liu C 2009 Decadal climatic variability, trends and future scenarios for the North China Plain *Journal of Climate* **22** 2111-2123

[5] Liu W, Fu G, Liu C, Song X and Ouyang R 2013 Projection of future rainfall for the NCP using two statistical downscaling models and its hydrological implications *Stochastic Environmental Research and Risk Assessment* **27** 1783-1797

[6] Guo Y, Li J and Li Y 2011 A time-scale decomposition approach to statistically downscale summer rainfall over North China *Journal of Climate* **25** 572-591

[7] Buonomo E, Jones R, Huntingford C and Hannaford J 2007 On the robustness of changes in extreme precipitation over Europe from two high resolution climate change simulations *Quarterly Journal of the Royal Meteorological Society* **133** 65-81

[8] Coordinated Regional Climate Downscaling Experiment. Region 7: East Asia. Available online: https://cordex.org/domains/region-7-east-asia/ (Accessed 27/05/2020)

[9] Asian Precipitation - Highly-Resolved Observational Data Integration Towards Evaluation. Available online: https://climatedataguide.ucar.edu/climate-data/aphrodite-asian-precipitation-highly-resolved-observational-data-integration-towards (Accessed 12/06/2020)
[10] National Centre for Atmospheric Research. China Monthly Station Precipitation and Temperature, 1951-2000. Available online: https://rda.ucar.edu/datasets/ds578.1/ (Accessed 12/06/2020)
[11] Wu D, Fang S, Li X, He D, Zhu Y, Yang Z, Xu J and Wu Y 2019 Spatial-temporal variation in irrigation water requirement for the winter wheat-summer maize rotation system since the 1980s on the North China Plain *Agricultural Water Management* **241** 78-86
[12] Thornthwaite C W 1948 An approach toward a rational classification of climate *Geographical Review* **38** 55-94
[13] Mann H 1945 Nonparametric tests against trend *Econometrica* **13** 245-259
[14] Kendall M 1975 *Rank Correlation Measures* (London: Charles Griffin)
[15] Tao F and Zhang Z 2010 Adaptation of maize production to climate change in North China Plain: quantify the relative contributions of adaptation options *European Journal of Agronomy* **33** 103-116
[16] Liu C, Zhang X and Zhang Y 2002 Determination of daily evaporation and evapotranspiration of winter wheat and maize by large-scale weighing lysimeter and micro-lysimeter *Agricultural and Forest Meteorology* **111** 109-120
[17] Gao X, Shi Y, Song R, Giorgi F, Wang Y and Zhang D 2008 Reduction of future monsoon precipitation over China: comparison between a high resolution RCM simulation and the driving GCM *Meteorology and Atmospheric Physics* **100** 73-86
[18] Hu S, Mo X, Lin Z and Liu S 2016 Adaptation of winter wheat to climate change in Huang-Huai-Hai Plain *Journal of Natural Resources* **31** 1892-1905

Acknowledgments
We acknowledge ViaX Education for supporting this research.

IOP Conf. Series: Earth and Environmental Science 1006 (2022) 012004 doi:10.1088/1755-1315/1006/1/012004

Possibility for Using Biochar for Sustainable Agriculture: a Case Study in Bulgaria

Rositsa Velichkova[1], Wandong Zheng[2], Detelin Markov[1], Iskra Simova[1], Ivan Ivanov[3], Ivan Denev[3], Martin Pushkarov[4]

1 Technical University of Sofia, Department of Hydroaerodynamics and Hydraulic machines, Centre for Research and Design in Human Comfort, Energy and Environment (CERDECEN), 1000 Sofia, Bulgaria
2 Tianjin University, School of Environmental Science & Engineering, 300457 Tianjin, China
3 Technical University of Sofia, Faculty of Engineering and Pedagogy of Sliven, 8800 Sliven, Bulgaria
4 Technical University of Sofia, College of Energy and Electronics (CEE), Department of Energy and Mechanical Engineering, 1000 Sofia, Bulgaria
E-mail: rvelichkova @tu-sofia.bg

Abstract. In both industrialized and developing countries, soil loss and degradation are observed at unprecedented rates and profound implications for soil ecosystem properties. In many regions, loss of soil productivity occurs despite the intensive use of agrochemicals, in parallel with the adverse environmental impact of soil and water resources. The soil types and the agro-ecological areas in Bulgaria are presented. Four samples of chernozem soil were collected and were investigated with and without biochar. The results obtained show the effect of adding biochar to the change of the pH of the soil

1. Introduction

The question of growing food in a changing climate is one of today's challenges. All people eat. Estimates show environmental changes could decrease wheat yields in non-industrial nations by 13%, and rice harvests could fall by 15% by 2050. In Africa, corn yields could fall by 10-20% for a similar period [1]. The Intergovernmental Panel on Climate Change has assessed that worldwide food creation could be decreased by up to 17% by 2100 as a result of harvest disappointments because of environmental change [2]. The Earth populace simultaneously is assessed to be 11.2 billion individuals in 2100 [3].

Agribusiness is important for the worldwide carbon crisis. Exploration shows that agribusiness is liable for 33% of the worldwide ozone harming substance outflows [4]. Food creation represents the largest part of all food-related greenhouse gas emissions– 86% [5]. Food growing is inevitably related to the factors that affect climate change.

A lot of research studies propose that biocoke may assume a part in supporting different practices to settle soil carbon. Biocoke stores roughly half of the carbon in the biomass changed over to pyrolysis, which can proceed as a steady wellspring of carbon in the soil for millennia [6]. The study in [7] has found that carbon diminished carbon mineralization and discharged between 64.9% and 68.8%, a phenomenon called the negative effect of priming [7]. This means that the carbon that is already in the soil is stabilized and retained in it.

It is hard to foresee the effect of biomass in the soil completely. The diverse types of biocoke delivered from various crude materials – crop residues, wood, excrement – appear to obtain unique nutritional and physical properties. They will affect the particular soil type distinctively. The temperature and pH of the biomass changed over into biocoke seem to affect its properties significantly.
The aim of the paper is to examine different types of soils from Bulgaria. Four samples of chernozem soil were collected and were investigated with and without biochar. The effect of adding biochar to the change of the pH of the soil is assessed.

2. Agro-Ecological Zoning in Bulgaria

Crop zoning aims to establish the most appropriate and correct relationship between: (i) the climatic and soil characteristics of the particular regions and even neighbour areas; (ii) the characteristic requirements of the crops to the environmental characteristics; (iii) the potential of the environment to provide the best possible conditions for the growth of the crops; (iv) the possible harvest, (v) the farms' capacity for both the production and use of agricultural products.
In other words, the main task of zoning crops, based on each region's soil, climatic and economic conditions, is to zoning crops to obtain the highest yields and highest productivity at a minimum unit cost of production for each individual holding or farm.
On the other hand, the zoning of agricultural crops must ensure the fullest and most rational use of land by each farm as the primary means of production and ensure maximum production from it and a continuous increase in its fertility, using the maximum degree of natural soil potential.
The rational zoning of crops aims not only to determine the most accurate specialization and organization of farms for each region through the maximum use of natural resources and labour resources of each region but also to solve the problem of the right combination of individual branches of agriculture by regions, the correct alternation of crops and the establishment of proper crop rotation.
A careful study of the influence of climate and soil on the development of agricultural crops and the size of their yields shows that not all elements of climate and especially soil are equally important for the success of the crop. There is always one main decisive factor, and others - secondary.
The analysis of the relationship between the development of crops and their yields and the individual soil and climatic elements determines the boundaries of each region or in other words – the agro-ecological region.
Each agro-ecological region is characterized by strictly defined and specific agro-climatic indicators for each crop. The agro-climatic indicators are a quantitative expression of the requirements of the respective crop to the soil conditions and climatic factors decisive for its yield.
The agro-ecological zoning is based mainly on the soil and climatic characteristics of each part of the country, as well as on the basic requirements of the crops to the soil regarding the main nutrients and their quantities, to the physical and water properties of the soil, to the reaction of the soil solution) and other.
The agro-ecological zoning divides the country's territory into seven main soil-geographical zones and the intra-zonal regions separated in each of them, based on the peculiarities of the soil cover and the bioclimatic factors.
The individual zones include the soil types in the country, and the agro-ecological areas belonging to them are shown in Fig. 1, where:

IOP Conf. Series: Earth and Environmental Science 1006 (2022) 012004 doi:10.1088/1755-1315/1006/1/012004

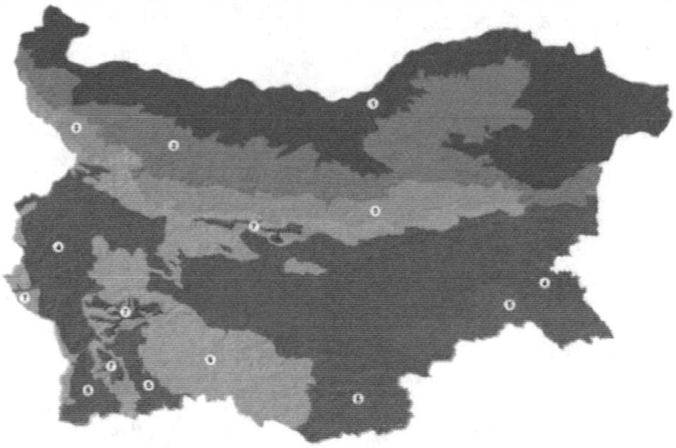

Figure 1. Map of different soil in Bulgaria

1. Chernozem region.
2. Grey-brown forest soil area.
3. Grey-brown podzolic soil region.
4. Chernozem region. resins and cinnamon forest soils.
5. A region of cinnamon forest soils in the transitional Mediterranean sub-zone.
6. An area of brown mountain forest soils.
7. An area of mountain-meadow soils.

3. Experiments

Samples of humus (chernozem) soil were taken from three places: slops of the Rila mountain, south-western region (near Blagoevgrad) and Sofia region. The fourth type was soil from the market. Samples of 100 g were made, measured on e KERN ABS 220-4N digital scale with an accuracy of 0.0001 g.

Figures 2-5 shows photos of the investigated soils, taken with LEVENHUK DTX 500 digital microscope.

Figure 2. Soil sample 1

IOP Conf. Series: Earth and Environmental Science 1006 (2022) 012004 doi:10.1088/1755-1315/1006/1/012004

Figure 3. Soil sample 2

Figure 4. Soil sample 3

Figure 5. Soil sample 4

The pH of the soil samples was measured with a pH meter for soil MW101-SOIL, with accuracy ±0.02. The experimental set-up is shown in Fig. 6.
Then new samples of mixed soil were made, consisted of 90 g soil and 10 g biochar. Figs. 7-10 show the microscopic pictures of the new samples. The pH of the enriched soil samples was measured.

Figure 6. Set-up for pH measurement

Figure 7. Picture of mixed soil 1: 90g soil 1+10 g biochar

Figure 8. Picture of mixed soil 2: 90g soil 1+10 g biochar

Figure 9. Picture of mixed soil 3: 90g soil 1+10 g biochar

Figure 10. Picture of mixed soil 4: 90g soil 1+10 g biochar

Table 1. pH results

	Sample 1-100gr soil	Sample 1-90gr soil+10 gr biochar
pH	*4,71*	*5,25*
	Sample 2-100gr soil	Sample 2-90gr soil+10 gr biochar
pH	*8,37*	*8,55*
	Sample 3-100gr soil	Sample 3-90gr soil+10 gr biochar
pH	*7,73*	*8,9*
	Sample 4-100gr soil	Sample 4-90gr soil+10 gr biochar
pH	*6,65*	*8,55*

Based on these results, a future experiment with the growth of tomatoes and cucumbers is planned. The Sample 4 soil will be used as in that case, the greatest increment of pH was obtained.

4. Conclusion

In the presented work, an experiment was made with four soil samples, first without impurities and then with 10% biochar. It can be seen that the pro addition of it increases the acidity of the soil. Based on these studies, a pilot selection of soil was made, which will be studied in real conditions. Hydrochar will also be included in future experiments.

5. References

[1] https://www.nature.com/news/one-third-of-our-greenhouse-gas-emissions-come-from-agriculture-1.11708

[2] https://www.ipcc.ch/pdf/assessment-report/ar5/wg2/drafts/fd/WGIIAR5-Chap7_FGDall.pdf

[3] http://www.un.org/sustainabledevelopment/blog/2017/06/world-population-projected-to-reach-9-8-billion-in-2050-and-11-2-billion-in-2100-says-un/

[4] https://www.nature.com/news/one-third-of-our-greenhouse-gas-emissions-come-from-agriculture-1.11708

[5] https://www.nature.com/news/one-third-of-our-greenhouse-gas-emissions-come-from-agriculture-1.11708

[6] http://www.css.cornell.edu/faculty/lehmann/publ/FrontiersEcolEnv%205,%20381-387,%202007%20Lehmann.pdf

[7] https://www.researchgate.net/publication/262568549_Biochar_suppressed_the_decomposition_of_organic_carbon_in_a_cultivated_sandy_loam_soil_A_negative_priming_effect

Acknowledgments

This study is part of the project „ Recycling of carbon fertilizer from straw for sustainable agriculture", Д01-62/18.03.2021, funded by the Bulgarian Science Fund of the Ministry of Education and Science.

IOP Conf. Series: Earth and Environmental Science 1006 (2022) 012005 doi:10.1088/1755-1315/1006/1/012005

A Comparative Study of Geometric and Exponential Laws in Modelling the Distribution of Daily Precipitation Durations

Mounia El Hafyani and Khalid El Himdi

Laboratory of Mathematics, Mohamed V University, Rabat Morocco
E-mail: mounia.elhafyani@gmail.com

Abstract. Wet and dry duration modelling is critical in engineering hydrology. The goal of this study is to model and analyze the wet and dry duration distributions. For this intent, daily rainfall data for Kenitra station were used on the period from 1967 to 2017. To represent the distribution of wet and dry durations, First-order Markov chain, Second-order Markov chain, and truncated negative binomial distribution are applied to represent the distribution of wet and dry durations. To assess the data adherence to the proposed models, the Chi-square and Kolmogorov-Smirnov tests have been used. The Akaike information criterion is applied to determine the most effective model distribution. We go further to investigate the distribution of the number of wet and dry days over a k-day period. This law is implemented using an algorithm based on conditional laws. This work is completed by comparing the calculated moments of the three estimated models to the observed moments of the number of wet/dry days over k consecutive days. The study demonstrates the effectiveness of our method for modelling wet and dry daily precipitation durations.

1. Introduction

Morocco is a country in the subtropical region of Northwest Africa. The geographical position offers the country a wide variety of bioclimatic; its climate is characterized by a striking spatial and temporal variability of precipitation. There are many nuances in its climate: The north is the Mediterranean, the west is oceanic, the interior is continental, and the south is the Saharan. The climate varies also according to the seasons; the majority of Moroccan regions are mountain parts which make them sensitive to the effect of global warming [1].

The modification of the precipitation profile is among the considerable preoccupations of climate change in North African, including Morocco. Precipitation has noted a significant reduction since 1960, [2]. Periods of drought are more present with destructive consequences on the economy of the country which has suffered from damaged flooding events causing serious losses and degradation. [3]. The tendency of precipitation and temperature has widely studies by researchers reported movement towards warmer and drier conditions [4] marked downwards trends. Indeed, semi-arid areas especially in the south and east are more vulnerable to climate change. A raise in the variability of annual rainfall, from the intervals of 1931–60 to 1961–90 is noted for the south of the Atlas Mountains [5].

This study is focused on the station of Kenitra, located in the north-western of Morocco at the extreme downstream of the Sebou basin; it's situated in a highly humid zone that contains many wetlands and rice farms. Its Mediterranean climate is characterized by the alternation of mild wet winters from October to April and dry and hot-dry summers from May to September [6]. However, the beginning of the end of the two seasons can be shifted by a few weeks.

Kenitra city is exposed to floods, frequently observed during rainy periods; they are generated by two different sources: the overflowing of Oued Sebou and floods from Oued Fouarat [7]. The opening of the station to the Atlantic Ocean has earned it interesting rainfall, with averages around 562 mm over the 50-year study, the station recorded a maximum value of 1130 mm during the year 1996 and a minimum value of 256 mm during the year 2015, the rainfall generally shows a peak during the months of November and December.

2. Importance of studying length of wet and dry spells

The statistical study of hydro-meteorological phenomena occupies an important place in several theoretical and practical fields. Across the weather variables that impact agriculture, rainfall is the most crucial of the meteorological elements. In the last few years, precipitation has had the most effect on environmental conditions and productive activities, for the reason of the fluctuations in the amount, time of occurrence, and duration. Structural changes of precipitation regimes were reported for continents [8].

The sequences of dry days are recurring events with some of the major huge worldwide effects. Dry spells have become one of the most important focuses since climatic models reported increasing incidence of summer droughts under global warming in zones where droughts were infrequent events [9]. These natural disasters have continuous repercussions on the economic and social components of the region where they happen; sometimes global assistance interventions are applied.

The sequence of rainy days can as well cause economic losses as it can influence the quality of agricultural products if it takes place during the crop periods [10]. It can also amplify the probability of flood events with serious repercussions/sequels on infrastructures such as road and rail networks, water processing, electricity supply, and barrage.

A daily wet/dry spell is defined as a period of multiple consecutive days with an amount higher/lower than a certain threshold; the wet spell is preceded and continued by a dry spell, a dry/wet spell with too small/large amount of precipitation could cause an extreme event [11].

2.1. Distributions used in literature

There is no consensus on the choice of the model to be used to adjust the daily rainfall sequences. However, using one distribution rather than another can significantly change the results and consequences of the analysis. It is therefore essential to represent the rainfall series by the best potential distribution. A large number of statistical distributions and parameter estimation methods are available and have been used, depending on their category, i.e. rainy or non-rainy.

Research on the estimation of a sequence of wet and dry spells has been well dealt with by a number of investigators, e.g [12, 13, 14]. The Markov-chain probability model was widely employed for stochastic modelling of the sequences of wet and dry days. Gabriel and Neumann [15] adopted the Markov chain to characterize the high persistence of dry days. This method was then used by numerous authors e.g [16], [17], and [18] they gave a good practice of this approach.

In more recent papers, the Markov chains approach was adopted by Junior et al. [19], who employed the daily rainfall to generate sequences of wet and dry days, [20] showed the Markovian model efficiency in estimating dryness persistence. Admasu et al. [21] adopted Markov chain analysis to model sequences of spells so as to comprehend the drought trend of the region. Back et al. used the Markov chain to represent the daily rainfall and estimate the number of dry and wet days for some areas of Santa Catarina [22]. However, the Markov chains are not entirely suitable for some regions, for example, the southern Iberian Peninsula [23]. Markov chains with different orders are as well applied [24]. The order kept in the analysis of the daily precipitations series was specified using the method of AIC (Akaike information criterion) and BIC. The latest investigations have favourably used BIC, which has been proved to be asymptotically unbiased [25] Negative binomial distribution was used by [20], and [26] to analyze the behavior of dry sequences. The truncated negative binomial gave good estimations, especially for long dry spells [27]. Some researchers compare multiple distributions for the same data,[28] applied the Geometric, Logarithmic, and Negative Binomial Truncated distributions to study the sequence of the probabilities of dry and wet days, Even if the negative binomial truncated is the best, the geometric distribution gives also good adjustments. For the same objective, Chowdhury and Beecham [27] analyze the group of geometric, GP, Logarithmic, Polya, and the negative binomial, the best results are achieved by geometric and composed geometric distributions.

3. Method and approach

We are working on the daily precipitation observed from the Kenitra station during the period "1967-2017". The objective of this study is to model the succession of days Jt according to their dry or wet character. A day is declared wet (resp dry) if the amount of precipitation is greater (resp less) than a defined α threshold. We thus define the state variable Jt which takes the value 1 (wet day) if Jt ≥ 0.1 and 0 (dry day) if Jt < 0.1. The spells were listed by months when a spell overlapped the end of the month, it was assigned to the month in which the last day arised.

4. Theory

4.1. The first order Markov chain

This is a particular model of the internal renewal process which corresponds to the case of constant persistence.

The probability law P_0^n, which represents the duration of a dry sequence, correspond to the waiting time for the first success in a series of independent Bernoulli tests with the same parameter $1-q^i$, it's a geometric law.

The form of the probability distribution function is: $p_n^i = \left(1-q^i\right)\cdot\left(q^i\right)^{n-1}$; For $n \geq 1$; $i = 0,1$

The parameter q^i is estimated by the maximum likelihood method on a sample $x_1^i, x_2^i, \ldots, x_k^i$ of lengths of sequences for the state i. $\hat{q}^i = 1-\left(1/\overline{x}^i\right)$, where $\overline{x}^i$ is the empirical mean

4.2. The second order Markov chain

This is a particular pattern of the alternating renewal process that corresponds to the case of constant persistence from the second day of the sequences.

The parameters of the law p_n^i are $\left(q_1^i, q_2^i\right)$. The form of the probability distribution function is:

$$p_n^i = \begin{cases} 1-q_n^i & \text{if } n=1 \\ q_1^i\cdot\left(1-q_2^i\right)\cdot\left(q_2^i\right)^{n-2} & \text{if } n \geq 2 \end{cases}$$

The parameters q_1^i and q_2^i are estimated by the maximum likelihood method on a sample $x_1^i, x_2^i, \ldots, x_k^i$ of lengths of sequences for the state i.

$q_1^i = k_2^i / k$; $q_2^i = 1-k_2^i / \left(k.\overline{x^i}-k\right)$ k_2^i Designates the number of sequences of lengths greater than or equal to 2

4.3. The truncated negative binomial law

Given h and d two positives real parameters. The law $\left(p_n, n \geq 1\right)$ is called BNT(h,d) if:

$$p_n^i = (1+d)^{(-h/d)} \cdot \frac{(h/d)_{n-1}}{(n-1)!} \cdot [d/(1+d)]^{n-1}$$

If $d/h > 1$ resp $(d/h < 1)$, the sequence $\left(q_n, n \geq 1\right)$ is monotonic increasing (resp decreasing) with limit $d/(d+1)$ and if $d/h = 1$ the sequence is constant and takes the value $d/(d+1)$.

The parameters (h,d) are estimated by the method of moments: $\hat{h} = \overline{x}-1$ And $\hat{d} = \left[s^2/(\overline{x}-1)\right]-1$

Where s^2 denotes the empirical variance of the sample $x_1,\dots,x_k$ $s^2 = (1/k)\sum_{i=1}^{k}\left(x_i - \overline{x}\right)^2$

5. Application and results

5.1. Monthly analysis of the frequency of the length of the wet/dry spells

Table 1. Fitted Models for Wet Spells

Months / Parameters		9	10	11	12	1	2	3	4	5	6	7	8
Mean		1,52	1,93	**2,71**	**2,72**	2,86	**2,65**	**2,70**	2,31	1,98	1,43	1,13	1,14
variance		0,69	1,59	5,72	4,83	7,06	5,58	4,22	2,62	1,94	0,48	0,12	0,12
Model1	**qi**	0,34	0,48	**0,63**	**0,63**	0,65	**0,62**	**0,63**	0,57	0,49	0,30	0,11	0,12
Model2	**qi1**	0,37	0,50	0,60	0,64	0,65	0,61	0,64	0,61	0,53	0,33	0,13	0,14
	qi2	0,30	0,47	0,65	0,63	0,65	0,63	0,62	0,54	0,46	0,23	0,00	0,00
Model3	**hi**	0,52	0,93	1,71	1,72	1,86	1,65	1,70	1,31	0,98	0,43	0,13	0,14
	di	0,31	0,71	2,35	1,80	2,79	2,37	1,49	1,00	0,98	0,11	-0,07	-0,11
	dhi	0,59	0,76	1,37	1,05	1,50	1,43	0,88	0,77	1,00	0,26	-0,53	-0,78

Table 2. Fitted Models for Dry Spells

Months / Parameters		9	10	11	12	1	2	3	4	5	6	7	8
Mean		31,70	10,11	**6,58**	**5,20**	**6,47**	**5,47**	**6,89**	5,77	7,94	13,66	28,88	52,20
variance		1281,85	303,29	67,58	37,67	66,49	34,81	88,04	38,18	93,74	140,54	453,24	794,65
Model1	**qi**	0,97	0,90	**0,85**	**0,81**	**0,85**	**0,82**	**0,85**	0,83	0,87	0,93	0,97	0,98
Model2	**qi1**	0,87	0,74	0,77	0,73	0,72	0,75	0,73	0,75	0,74	0,92	1,00	1,00
	qi2	0,97	0,92	0,86	0,83	0,87	0,83	0,88	0,84	0,89	0,93	0,96	0,98
Model3	**hi**	30,70	9,11	5,58	4,20	5,47	4,47	5,89	4,77	6,94	12,66	27,88	51,20
	di	40,76	32,31	11,12	7,97	11,15	6,78	13,93	7,01	12,51	10,10	15,26	14,52
	dhi	1,33	3,55	1,99	1,90	2,04	1,52	2,36	1,47	1,80	0,80	0,55	0,28

Parameter and probability estimation: Table1 and Table2 displays, per month, the mean length of dry and wet spells, and the parameters associated with the three models fitted. The average length of wet spells ranged from 1.13 to 2.86 days while that of dry spells varied from 5.20 to 5.20 days; the variability of dry spells length is significant.

The selection of the best probability model for every month: The model performance was assessed for its capacity to explain the behavior of the distribution of wet/dry spells in Kenitra. This was performed by using the goodness-of-fit tests that summarize the difference between the observed and fitted values and suggest the most representable model for observed data. the chi-square goodness of fit test and Kolmogorov-Smirnov test has been employed, with alpha= 0.05.

The Kolmogorov-Smirnov test for goodness of fit is one of the most robust non-parametric tests for differences among two cumulative frequency distributions of the observed and expected ones. The probability model which related to the minimum value of Dmax constitutes the best fit model in performing the distribution of wet/dry spells. The Chi-square test is critical to very small cell frequencies, it's inadequate when estimated frequencies are less than 5 in 20 percent of the cells. The Chi-square test requires grouping of the data for classes with low counts, while KS test deals with individual observation singly begetting to no loss of information in grouping whereas loss of information in chi-square procedure is wide.

Table 3. Model selection per month for wet period

Month	Model	Validation Khi2	pvalue khi2	D_KS	D0,05	AIC	BIC	Valid model	Best model AIC	Best model KHI2	Best model KS	Best Model
1	MC1	6,4	0,59	0,039	0,1	96,41	104,72	MC1,MC2,BN	BN	MC1	MC1	MC1
	MC2		0,42	0,041		98,29	114,90					
	BN		0,08	0,064		96,12						
2	MC1	4,0	0,07	0,067	0,1	88,26	96,59	MC1,MC2,BN	BN	MC1	MC1	MC1
	MC2		0,03	0,074		89,89	106,55					
	BN		0,02	0,095		87,99						
3	MC1	5,8	0,76	0,027	0,1	85,41	93,70	MC1,MC2,BN	MC1	MC1	BN	MC1
	MC2		0,65	0,026		87,79	104,38					
	BN		0,68	0,024		88,53						
4	MC1	4,3	0,49	0,028	0,1	58,23	66,68	MC1,MC2,BN	MC1	MC2 & BN	BN	BN
	MC2		0,85	0,011		61,36	78,26					
	BN		0,8	0,01		61,27						
5	MC1	4,7	0,19	0,064	0,11	53,95	61,97	MC1,MC2,BN	MC1	MC2	MC2	MC2
	MC2		0,45	0,023		57,81	73,86					
	BN		0,07	0,068		55,70						
6	MC1	1,4	0	0,035	0,16	19,22	25,68	--	--	--	--	
	MC2		NA	0,013		21,73	34,66					
	BN		NA	0,009		21,78						
7	MC1	12,5	NA	0,026	0,34	6,87	10,41	--	--	--	--	
	MC2		NA	0		8,43	15,52					
	BN		NA	0,007		8,54						
8	MC1	13,8	NA	0,031	0,25	6,74	11,47	--	--	--	--	
	MC2		NA	0		8,26	17,73					
	BN		NA	0,004		8,32						
9	MC1	3,6	0,29	0,048	0,15	18,18	25,05	MC1,MC2,BN	MC1	MC1	BN	MC1
	MC2		0	0,037		20,32	34,04					
	BN		0	0,036		20,20						
10	MC1	1,8	0,91	0,009	0,11	41,83	50,02	MC1,MC2,BN	MC1	MC1	MC1 & BN	MC1
	MC2		0,83	0,01		44,15	60,53					
	BN		0,81	0,009		44,46						
11	MC1	7,0	0,2	0,032	0,1	87,03	95,48	MC1,MC2,BN	MC1	MC1 & BN	MC1	MC1
	MC2		0,15	0,048		87,91	104,81					
	BN		0,22	0,045		87,36						
12	MC1	5,6	0,75	0,023	0,1	65,78	74,32	MC1,MC2,BN	MC1	MC1	MC1 & MC2	MC1
	MC2		0,62	0,023		67,78	84,87					
	BN		0,57	0,027		67,56						

For dry spells, from April to October, we have found that more than 20 percent of cells' expected frequencies are less than 5, the khi2 test is thus not applicable. the verification of khi2 test applicability is situated in the first column (Table 3 and Table 4)

The choice of the best model is based on two criteria: the first one is according to Akaike's Information Criterion (AIC), while the second one is the selection of the model that has the most significance using the khi2 and KS tests. If the first condition is filled but not the second one, the following model with the minimum value of AIC which satisfies the second condition is then privileged.

For wet spells, the findings in Table 5 show that the three models are significant for all months. The MC1 is chosen for the majority of months except for April and Mai which are respectively fitted by BN and MC2. This may be due to a prolonged period of dry spells and a small period of wet spells that regularly arise in these months than in the other months. The results in Table 6 show best results for second-order Markov chain and truncated binomial negative which are more efficient in fitting observed data for dry spells. The first-order Markov chain is not adequate for January, Mars, Mai, October, and November.

Table 4. Model selection per month for dry period

Month	Model	Validation Khi2	pvalue khi2	D_KS	D0,05	AIC	BIC	Valid model	Best model			Best Model
									AIC	KHI2	KS	
	MC1		0	0,13		220,52	228,78					
1	MC2	18,3	0,3	0,043	0,1	169,72	186,24	MC2, BN	BN	BN	BN	BN
	BN		0,61	0,028		224,57						
	MC1		0,01	0,074		169,66	178,05					
2	MC2	13,9	0,06	0,055	0,1	136,27	153,04	MC1,MC2, BN	BN	BN	BN	BN
	BN		0,23	0,03		171,64						
	MC1		0,02	0,133		205,53	213,82					
3	MC2	18,7	0,61	0,044	0,1	153,48	170,04	MC2,BN	MC2	MC2	BN	MC2
	BN		0,2	0,018		214,52						
	MC1		0,31	0,074		205,53	213,95					
4	MC2	16,9	0,26	0,024	0,1	154,26	171,10	MC1,MC2,BN	MC2	MC1	MC2	MC2
	BN		0,15	0,027		205,55						
	MC1		0	0,123		229,62	237,55					
5	MC2	33,6	0,05	0,041	0,11	138,13	153,98	MC2,BN	MC2	--	BN	MC2
	BN		0,15	0,025		236,48						
	MC1		0,19	0,031		234,79	241,13					
6	MC2	83,1	0,13	0,031	0,17	95,65	108,35	MC1,MC2,BN	MC2	--	MC1 & MC2	MC2
	BN		0,09	0,036		233,96						
	MC1		0	0,039		126,43	130,09					
7	MC2	100,0	NA	0,059	0,33	Inf	Inf		--	--	--	
	BN		NA	0,053		138,95						
	MC1		0	0,054		229,46	234,26					
8	MC2	100,0	0	0,054	0,25	Inf	Inf		--	--	--	
	BN		0	0,064		300,81						
	MC1		0	0,101		366,93	373,77					
9	MC2	68,7	0	0,04	0,15	101,89	115,56	MC1,MC2,BN	MC2	--	MC2	MC2
	BN		0,01	0,072		375,38						
	MC1		0,02	0,166		275,12	283,41					
10	MC2	23,4	0,03	0,047	0,1	155,44	172,01	MC2,BN	MC2		BN	MC2
	BN		0	0,033		304,26						
	MC1		0	0,099		221,50	230,01					
11	MC2	13,0	0,02	0,064	0,1	172,74	189,77	MC2,BN	MC2	BN & MC2	BN	BN & MC2
	BN		0,02	0,041		225,51						
	MC1		0,38	0,06		183,60	192,14					
12	MC2	9,7	0,52	0,049	0,1	147,64	164,74	MC1,MC2,BN	MC2	MC2	BN	MC2
	BN		0,4	0,033		185,15						

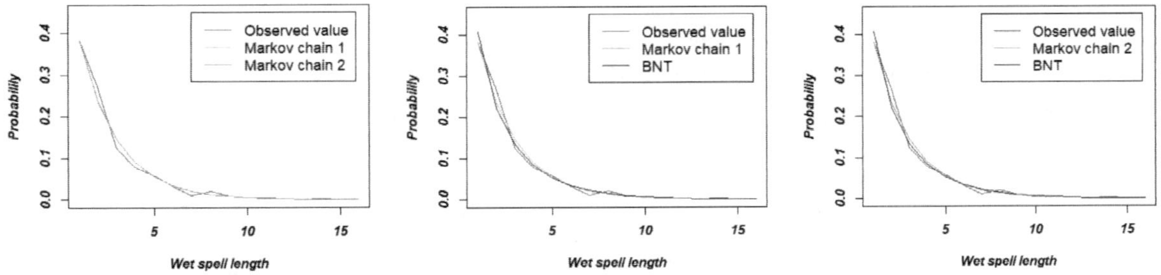

Figure 1. Observed and fitted distributions to wet spells for wet period

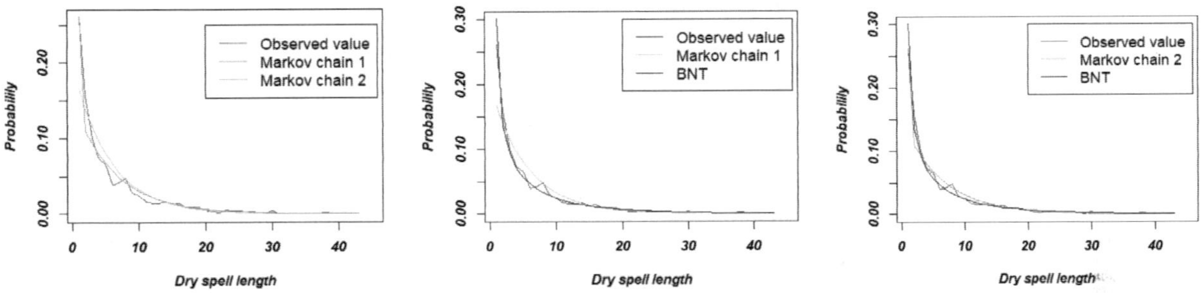

Figure 2. Observed and fitted distributions to dry spells for wet period

5.2. Analysis of the wet period

The objective of this section is to analyze the characteristic of the wet period and select the most adequate model for it. It lasts five months extending from November to April, the mean length of wet spells is close to 2,63 while the mean length of dry spells is close to 6,01. It is shown in Figure 1 that BNT's estimation is better than MC1 and MC2 which have similar estimations. For dry spells Figure 2 shows also that BNT is the most appropriate model for the data. By applying the AIC criteria, khi2 and KS tests, the selected Model is BNT for wet spells and dry spells.

Table 5. Model Selection for wet spells (wet period)

Model	Khi2 Test		KS Test		AIC	BIC	Best Model
	Statistic	Pvalue	D	D_0,05			
Markov1	25,58	0	0,03		147,91	159,90	
Markov2	25,55	0	0,03	0,041	149,79	173,76	BNT
BNT	24,78	0	0,05		145,05	--	

Table 6. Model Selection for Dry spells (wet period)

Model	Khi2 Test		KS Test		AIC	BIC	Best Model
	Statistic	Pvalue	D	D_0,05			
Markov1	135,66	0	0,1		484,94	496,93	
Markov2	56,82	0	0,05	0,041	388,07	412,05	BNT
BNT	29,68	0,13	0,02		445,99	--	

5.3. The law of the number of dry days among k consecutive days, for wet period

In this part, the distribution of the counting variable S10 is studied; it indicates the number of dry days among 10 consecutive days. For the three models. The closest estimates are determined by BNT model. These results are confirmed by Table 7 and Figure 3.

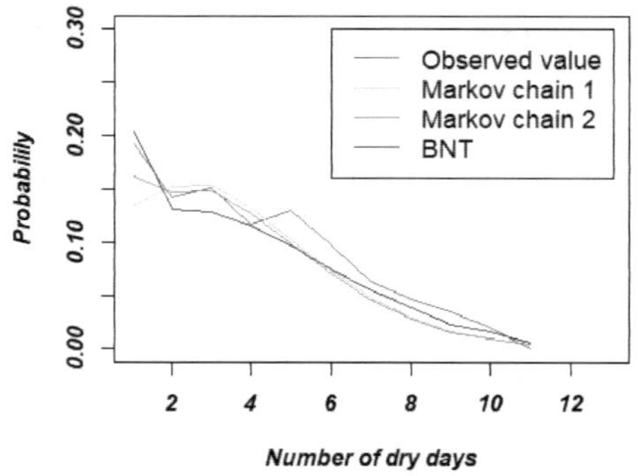

Figure 3. Number of dry days among k consecutive days

Table 7. Model selection for S010 distribution

Distribution	Khi2 Test		KS Test		AIC	Best Model
	Statistic	Pvalue	D	D_0,05		
Markov1	90,901	0	0,056		56,44	
Markov2	111,736	0	0,032	0,4457	58,82	BNT
BNT	82,031	0	0,034		56,73	

6. Conclusion

In this study, we examine the best model for fitting the distribution of the frequency of wet/dry spells at Kenitra's station. The most frequent model is selected as the most adequate for the distribution of dry and wet periods. Simulation of the frequencies of wet/dry spells per month showed that the MC model provided the closest estimates to the empirical frequencies of wet spells while best results for second-order Markov chain and truncated binomial negative for dry spells. For the wet period, the selected Model is BNT. The closest estimates are also determined by BNT model for the distribution of the number of dry days among 10 consecutive days.
Taking into account other factors such as wind flow patterns, temperature, and other weather factors, the study of the distribution of wet and dry periods may improve the understanding of hydrology, water resources management, and meteorology.

7. References

[1] Beniston, M. et al., 1996. Impacts of climate change on mountain regions. Cambridge, Cambridge University Press

[2] Knippertz, P., Christoph, M., Speth, P., 2003. Long-term precipitation variability in Morocco and the link to the large-scale circulation in recent and future climates. Meteorology and Atmospheric Physics 83, 67–88

[3] Mansouri, B., 2017. IMPACT OF DROUGHT AND FISCAL POLICY ON PRIVATE CONSUMPTION, PRIVATE INVESTMENT AND ECONOMIC GROWTH IN MOROCCO: AN EMPIRICAL ANALYSIS

[4] Driouech, F., Stafi, H., Khouakhi, A., Moutia, S., Badi, W., ElRhaz, K., Chehbouni, A., 2021. Recent observed country-wide climate trends in Morocco. International Journal of Climatology 41, E855–E874
[5] Hulme, M., 1992. Rainfall changes in Africa: 1931–1960 to 1961–1990. International Journal of Climatology 12, 685–699
[6] Calistri, P., Ippoliti, C., Candeloro, L., Benjelloun, A., El Harrak, M., Bouchra, B., Danzetta, M.L., Di Sabatino, D., Conte, A., 2013. Analysis of climatic and environmental variables associated with the occurrence of West Nile virus in Morocco. Preventive Veterinary Medicine 110, 549-553
[7] Bourak, A., Midaoui, A., Lahrach, A., Chaouni, A.-A., El Arrim, A., 2017. Modélisation hydraulique du système Sebou-Fouarat, ville de Kenitra, Maroc - Cas des inondations de 2010. European Scientific Journal.
[8] Groisman, P.Y., Knight, R.W., Karl, T.R., 2012. Changes in Intense Precipitation over the Central United States. Journal of Hydrometeorology 13, 47–66
[9] Caloiero, T., Coscarelli, R., Ferrari, E., Sirangelo, B., 2015. Analysis of Dry Spells in Southern Italy (Calabria). Water 7, 3009–3023
[10] Smiderle, O.J., Dias, C.T. dos S., 2011. Época de colheita e armazenamento de sementes de arroz produzidas no cerrado de Roraima. REVISTA AGRO@MBIENTE ON-LINE 5, 12–17.
[11] Gaume, E. et al., 2009. A compilation of data on European flash floods. Journal of Hydrology 367, 70–78.
[12] Williams, C.B., 1952. Sequences of wet and of dry days considered in relation to the logarithmic series. Quarterly Journal of the Royal Meteorological Society 78, 91–96
[13] Llano, M.P., Penalba, O.C., 2011. A climatic analysis of dry sequences in Argentina. International Journal of Climatology 31, 504–513
[14] Longley, R.W., 1953. The length of dry and wet periods. Quarterly Journal of the Royal Meteorological Society 79, 520–527
[15] Gabriel, K.R., Neumann, J., 1962. A Markov chain model for daily rainfall occurrence at Tel Aviv. Quarterly Journal of the Royal Meteorological Society 88, 90–95
[16] Caskey, J.E., 1963. A MARKOV CHAIN MODEL FOR THE PROBABILITY OF PRECIPITATION OCCURRENCE IN INTERVALS OF VARIOUS LENGTH. Monthly Weather Review 91, 298–301
[17] GALLOY, E., MARTIN, S., LEBRETON, A., 1982. Analyse des séquences de jours secs consécutifs. Application à 31 postes du réseau météorologique français. Analyse des séquences de jours secs consécutifs. Application à 31 postes du réseau météorologique français 6, 5–24
[18] Moon, S.-E., Ryoo, S.-B., Kwon, J.-G., 1994. A Markov chain model for daily precipitation occurrence in South Korea. International Journal of Climatology 14, 1009–1016
[19] Andrade Júnior, A.S. de, Frizzone, J.A., Sentelhas, P.C., 2001. Daily rainfall simulation in Parnaíba and Teresina - Brazil, using a spreadsheet. Revista Brasileira de Engenharia Agrícola e Ambiental 5, 271–278
[20] Anagnostopoulou, C., Maheras, P., Karacostas, T., Vafiadis, M., 2003. Spatial and temporal analysis of dry spells in Greece. Theoretical and Applied Climatology 74, 77–91
[21] Admasu, W., Tadesse, K., Yemenu, F., Abdulkadir, B., 2014. Markov chain analysis of dry, wet weeks and statistical analysis of weekly rainfall for agricultural planning at Dhera, Central Rift Valley Region of Ethiopia. AJAR 9, 2205–2213
[22] Back, Á., Della, J.P., Valvassor, I.M.L., Denski, A.P.N., 2014. Modelos para geração de séries sintéticas de precipitação diária. XII Simpósio Ítalo Brasileiro de Engenharia Sanitária e Ambiental, ABES, Natal 1–10
[23] Conesa Garcia, C., Martin-Vide, J., 1993. Analyse par la chaîne de Markov de la sécheresse dans le sud-est de l'Espagne. Sécheresse (Montrouge) 4, 123–129
[24] Alam, N.M., Sharma, G.C., Jana, C., Sharma, N.K., Mishra, P.K., Adhikary, P., Raizada, A., 2015. Probabilistic drought analysis of weekly rainfall data using Markov chain model. Journal of Reliability and Statistical Studies 105–114

[25] Schoof, J.T., Pryor, S.C., 2008. On the Proper Order of Markov Chain Model for Daily Precipitation Occurrence in the Contiguous United States. Journal of Applied Meteorology and Climatology 47, 2477–2486
[26] Mahmoudi, P., Parvin, N., Rezaei, J., 2013. Regionalization of Iran based on the length of dry spells. Journal of Arid Regions Geographics Studies 4, 85–106
[27] Chowdhury, R.K., Beecham, S., 2013. Characterization of rainfall spells for urban water management. International Journal of Climatology 33, 959–967
[28] Assis, F.N. de, 1996. Aplicações de estatistica a climatologia. UFPel, Ed. Universitaria

IOP Conf. Series: Earth and Environmental Science 1006 (2022) 012006 doi:10.1088/1755-1315/1006/1/012006

GIS Mapping of Salt Farms and Salt Supply Chain Actors in the Visayas and Mindanao, Philippines

G J C Bartolome[1], J P Magpantay[1], J A Delos Reyes[2], A T Lat[2], T J I Reodica[2], and C J B Manalo[2]

1 Cavite State University, Cavite, Philippines 4106
2 University of the Philippines Los Baños, Laguna, Philippines 4031
E-mail: gjcbartolome@cvsu.edu.ph

Abstract. The 2010 survey of the Nutrition Council of the Philippines (NCP) revealed dramatic changes in the level of participation of the salt supply chain actors and their production, processing, and marketing activities. Among the industry's biggest challenges are: salt remains excluded from the list of priority commodities of the government; the changes in weather patterns due to climate change which adversely impact the production cycle; prominence of traditional methods of production; and low productivity of local farms. While remote sensing technologies including geographic information system (GIS) mapping have been widely adopted for the improvement of agricultural productivity in the Philippines, use of such in research that focuses on salt production and supply chain is still very limited. GIS maps are perceived as effective information library and decision support tool that can help the supply chain of the neglected salt industry. This study focused on the mapping of salt farms and their supply chain actors from selected salt production sites in the Visayas and Mindanao islands. Eleven provinces were included in the surveys of market intermediaries and end-consumers who were traced from salt producers. Their geographic coordinates were recorded and satellite and Google Earth images of the supply chain actors were digitized in ArcGIS 10.5 software to generate the maps, physical characteristics of which were examined for possible patterns. Results show that in Mindanao, the market areas (or the salt traders) are concentrated in Zamboanga del Sur particularly in Aurora and Zamboanga City, but the producers are in Alubijid and El Salvador City, Misamis Oriental. In the Visayas, traders are more scattered and are greater in number than in Mindanao.

1. Introduction

The significance of salt in many of the world's economies can never be overemphasized as it is an important mineral, preservative, and ingredient in cooking. In addition, the bulk of salt produced worldwide goes to the industry for the manufacture of heavy chemicals, ceramics, glasses, textiles, and metals. It is also an extremely essential compound for proper nutrition and function of the human body by helping maintain internal balance, normal physiological activities, and addressing medical concerns like sore throat, toothaches, and digestion problems via intake of saline solutions [1-2].

Traditionally, salt is being produced from a variety of sources including salt rock, brine springs, and sea water. In other areas with rich salt resources, people developed ways to exploit them and by far, obtain salt through trade [3]. In most areas, salt is normally produced by first stabilizing seawater to remove impurities, followed by integrated processes of evaporation, concentration, and crystallization techniques to harvest the salt product [4]. In the Philippines, these processes are accomplished mostly using solar energy and the production of salt begins with the collection of seawater from the sea to shallow ponds. The ponds help concentrate the seawater by evaporation and crystallization while taking

IOP Conf. Series: Earth and Environmental Science 1006 (2022) 012006 doi:10.1088/1755-1315/1006/1/012006

advantage of the tropical climate and abundant sunlight. Another method of producing salt is by cooking in which seawater is boiled in large pans to evaporate excess moisture and eventually form salt crystals. Salt produced by cooking is finer in texture and whiter in color compared with granulated and off-white crystals from solar evaporation pans [5].

Worldwide, the Philippines ranked 28th among 91 countries in terms of the production of salt. In 2019, salt production reached 1,147.97 thousand metric tons (TMTs) with an average annual growth rate of 3.0%. China with 63,603.9 TMTs topped the global production of salt, followed by United States, India, Chile, and Australia. The Philippines was overtaken by Austria which is on the 27th with 1,206.58 TMTs and Italy at 1,107.72 TMTs [6]. Despite the recognized significance of salt in various industrial activities and its known domestic value, the salt industry is not included among the priority commodities of the country. In a report by the Pacific Farms, Inc. in 2009, the annual salt requirements of the Philippines reached 590,000 MTs. To address this demand during this period, about 400,000 MTs were imported from Australia and 22,000 MT from China [7]. The changes in seasonal patterns due to climate change and the reliance of salt producers on age-old production methods are the main challenges that significantly impact level of production of salt [8-11]. Meanwhile, another issue in agriculture development is the need to increase productivity, employment, and income of poor segments of the agriculture sector.

Geospatial technology such as the GIS mapping can help in creating dynamic and competitive agriculture support systems for data-driven decision making [12-13]. Remote sensing and GIS mapping have been known to be effective in estimating land suitability for agricultural production, determination of crop yield, soil nutrient distribution and pest management to improve agricultural productivity [14-18].

The survey of salt producers, traders, and consumers conducted by the Nutrition Council of the Philippines (NCP) in 2010 revealed evident changes in the salt industry profile in terms of the participation of supply chain actors, salt production, processing, and marketing activities. Despite the frontiers in remote sensing and GIS mapping to improve agricultural productivity in the Philippines, there is still a very limited number of researches that present local information related to the salt production. Thus, this study focused on the mapping of the salt farms and supply chain actors from selected salt production sites in Visayas and Mindanao islands. These maps are perceived as an effective information library and decision tool to support the initiatives of all stakeholders in reviving the Philippines' salt industry.

2. Methodology

2.1 Selection of study sites

Considering the temporal limitations that greatly influence the annual production of salt, priority was given to the provinces in the Philippines that produced salt in 2018. Eleven study sites were identified from the provinces of Iloilo (3), Antique (2), Guimaras (1), and Negros Occidental (3) in the Visayas island, and Misamis Oriental (1) and Zamboanga City (1) in Mindanao. Their geographic coordinates consisting of the latitude, longitude, and elevation were obtained using a Ashtech MobileMapper 10 handheld Global Positioning System (GPS) navigator. The boundaries of the salt farms were surveyed, and the waypoint coordinates were recorded for use in the generation of actual salt field maps.

2.2 Survey of the salt supply chain actors

A total of 376 supply chain actors composed of 149 salt producers, 70 intermediaries, 6 institutional buyers, and 151 household consumers were interviewed for the collection of the salt production and marketing data. Salt producers in each study site were chosen using simple random sampling if the population per area is >30, otherwise, complete enumeration was performed. For the intermediaries, the number of participants was determined until 80% of the total volume produced by the salt farmers have been traced. Lastly, for the consumers, a sample size of 151 was determined based on the current population considering a 50% response rate and a design effect of 1.5 (for risk factors).

Data collected include volume of salt production per municipality (in kg/ha), distribution of salt producers per type of production method (solar or cooking), distribution of salt producers per type of available technology being used (brick/tiles, clear PET film, black PET film, bamboo, cooking pan), production activities, and marketing practices.

2.3 GIS Mapping of the salt farms and salt supply chain actors

Geographic information systems (GIS) and remote sensing (RS) technologies were used in mapping the locations of the salt farms. An on-field mapping using an Ashtech MobileMapper 10 GPS navigator and a GIS-based validation technique were applied to manually digitize and determine the size of the selected salt farms. The GPS points collected from the field were saved in the GPS device and recorded manually. Meanwhile, with the consent of the salt producers, traders, and consumers, their approximate locations were obtained during the interviews. Participants' location was presented in the map by province and island. All the data were encoded in spreadsheets which contained the geographic locations in the form of X-Y coordinates. These coordinates were added to a shapefile using ArcGIS 10.5 software to create the map. A Landsat image and Google Earth image were used to overlay the locations of selected farms in the map and the approximate locations of the participants.

3. Results and Discussion

3.1 Salt production areas and farming technologies in Visayas and Mindanao

The locations of the salt production areas in the Visayas and Mindanao islands are shown in Figures 1 and 2. Out of 149 salt farms, only one farm produced salt using the cooking method while the rest of the producers employed solar evaporation technique. Most salt producers in the Visayas can be found in the provinces of Negros Occidental (24.8%) and Guimaras (20.1%). In the Mindanao region, salt farms are mostly situated in Misamis Oriental (28.9%).

The practice of salt-making in the Philippines is dominated by solar evaporation method (Table 1). Production of salt is carried out during the summer months between November to April, when there is abundant sunlight and weather condition is favourable to allow the formation of salt crystals in especially built salt ponds. The salt ponds, where the seawater is allowed to surge, circulated, and concentrated, are lined with various flooring materials including transparent polyethylene plastic (71.1%), black polyethylene plastic (13.4%) and brick tiles (6.7%). In some salt farms, seawater is moved from the source (sea) to the salt ponds using water pumps (powered by electricity or by wind energy) such as in Alubijid, Misamis Oriental and in Escalante, Pulupandan, and Bago cities in Negros Occidental. Some salt farms have also integrated filtration box systems to improve the quality of effluent from the salt ponds similar to the case of El Salvador City in Misamis Oriental. The filtration system consists mainly of sandy soils enclosed in a box made of bamboo and lined with polypropylene plastic sacks. On the other hand, to produce artisanal salt, some salt producers modified this technique using bamboo halves as evaporation vessel (8.1%). The effluent which is manually loaded in the bamboo halves is produced from the filtrates of sand saturated with seawater and filtered twice using an improvised filtering system. Meanwhile, with the cooking method, seawater is concentrated in a water tank and the stock is evaporated in a large cooking vessel by application of heat. Rice hull is commonly used as a heat source during cooking.

3.2 Quality of salt

Results of interviews showed that there are three classifications of salts that are being produced and traded within the supply chain (Table 2). Of the total 1,628,352.00 kg of salt produced in 2018, 51.97% is accounted for by Class A from the Visayas and Mindanao. Class A salt is characterized by the whitest crystals among the three classifications and is commonly used for home consumption. On the other hand, Class B and Class C constitute the 47.51% and 0.51% of the total production, respectively. Class B is distinguished by semi-white crystals and Class C by brownish color with visible impurities. Both classes are commonly used for industrial applications. Meanwhile, in Mindanao, Class B salt is used for the production of dried fish and as cleaning agent in water-refilling station. Class C commands the lowest prices among them.

IOP Conf. Series: Earth and Environmental Science 1006 (2022) 012006 doi:10.1088/1755-1315/1006/1/012006

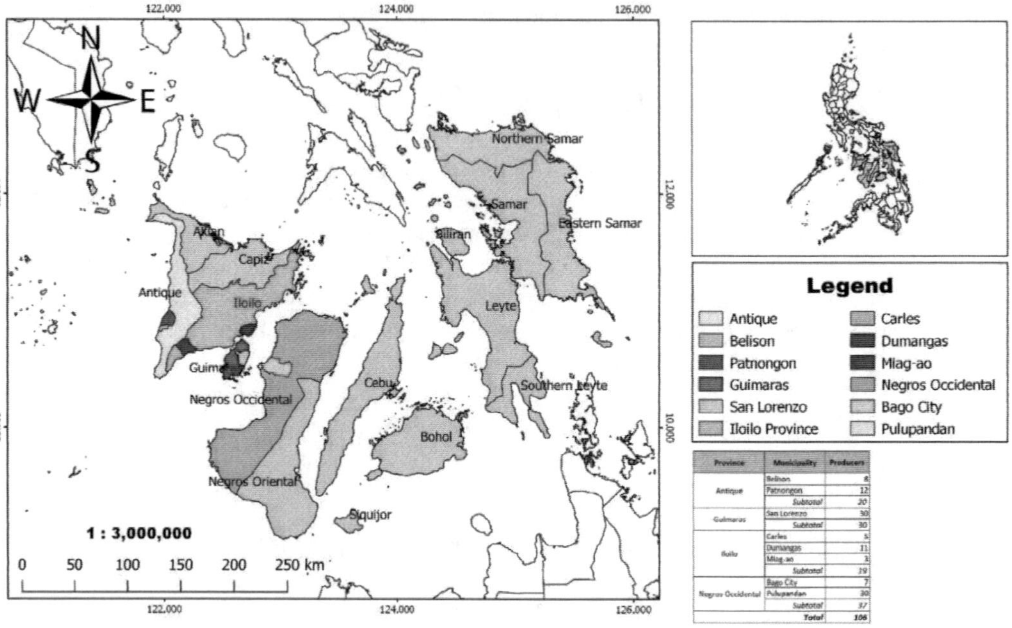

Province	Municipality	Producers
Antique	Belison	8
	Patnongon	12
	Subtotal	*20*
Guimaras	San Lorenzo	30
	Subtotal	*30*
Iloilo	Carles	5
	Dumangas	11
	Miag-ao	3
	Subtotal	*19*
Negros Occidental	Bago City	7
	Pulupandan	30
	Subtotal	*37*
	Total	***106***

Figure 1. Location of salt farms in the Visayas region

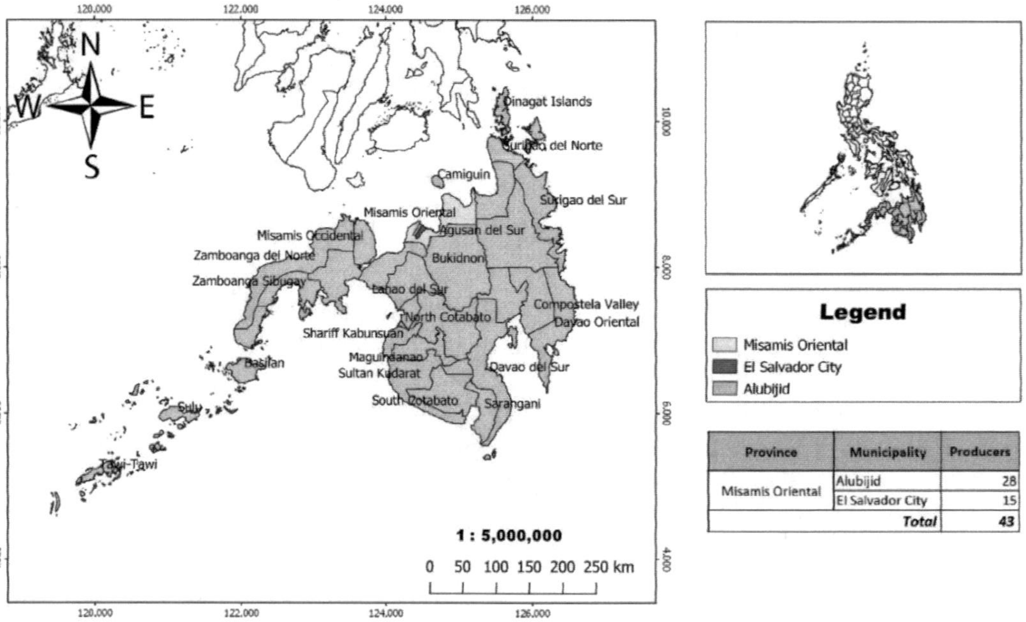

Province	Municipality	Producers
Misamis Oriental	Alubijid	28
	El Salvador City	15
	Total	***43***

Figure 2. Location of salt farms in the Mindanao region

3.3 Mapping of salt traders and consumers

Market intermediaries or traders refer to the market players who perform one or a combination of marketing practices in salt trading. These intermediaries can be classified into four categories who control the prices of the salt depending on the practices that they perform including value-addition activities. Among the intermediaries surveyed in the Visayas (Figure 3) and Mindanao (Figure 4) regions, the retailers were found to be the highest in terms of number (42.9%), followed by wholesaler-retailer (25.7%), assembler-wholesaler (22.9%), and wholesalers (8.6%).

Table 1. Distribution of salt producers by type of materials used in crystallization ponds versus cooking method (n=149)

Production area	Number of producers	Flooring materials in crystallization ponds								Cooking method		Total	
		Brick/tiles		Clear PE* plastic		Black PE plastic		Bamboo					
		No.	%	No.	%	No.	%	No.	%	No.	%	No.	%
Visayas													
Iloilo													
Carles	5	-	-	5	100	-	-	-	-	-	-	5	100
Dumangas	11	1	9	10	91	-	-	-	-	-	-	11	100
Miagao	3	-	-	-	-	-		3	100	-	-	3	100
Antique													
Belison	8	-	-	4	50	-	-	3	38	1	12	8	100
Patnongon	12	-	-	6	50	-	-	6	50	-	-	12	100
Guimaras													
San Lorenzo	30	-	-	30	100	-	-	-	-	-	-	30	100
Negros Occidental													
Bago City	7	-	-	7	100	-	-	-	-	-	-	7	100
Escalante	2	2	100	-	-	-	-	-	-	-	-	2	100
Pulupandan	30	-	-	30	00	-	-	-	-	-	-	30	100
Mindanao													
Misamis Oriental													
Alubijid	27	7	26	-	-	20	74	-	-	-	-	27	100
El Salvador City	15	-	-	15	00	-	-	-	-	-	-	15	100

**Polyethylene*

Table 2. Characteristics of salt and salt yield obtained from the study sites in Visayas and Mindanao

Location	Salt classification			Purity (% NaCl)	Color	Yield (kg/ha)
	Class A	Class B	Class C			
Visayas						
Iloilo						
Dumangas	✓	✓		67.26	White	8,333.33
Miag-ao	✓			68.87	Very light brown	32,000.00
Carles	✓	✓		92.63 67.00	White	36,111.11
Antique						
Patnongon	✓			72.08	Very light brown	312,500.00
Belison	✓			87.29 (bamboo) 74.48 (clear PE) 71.28 (cooking)	Very light brown White White	269,228.17
Guimaras						
San Lorenzo	✓	✓		89.88	White	16,667.67

Negros Occidental						
Escalante City		✓		91.82	White	21,093.75
Bago City	✓	✓		83.62	White	10,416.67
Pulupandan	✓	✓		77.08	White	13,888.89
Mindanao						
Misamis Oriental						
Alubijid	✓	✓	✓	84.32 (bricks) 81.07 (black PE)	White Slightly discolored	72,222.22
El Salvador City	✓			94.60	White	8,333.33
Total yield (Kg)	846,327.50	773,695.00	8,330.00			1,628,352.20

Source: Salt yield data (2019-2020)

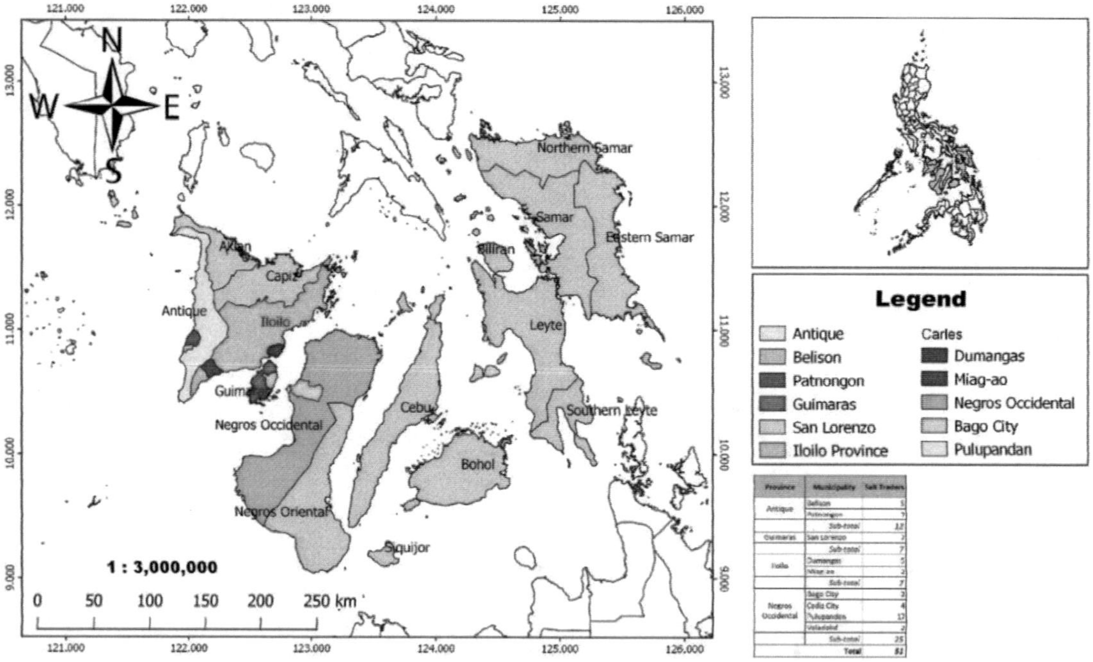

Figure 3. Location of market intermediaries in the Visayas region

The *viajeros* or assembler-wholesaler were traced in Negros Occidental (6%), Guimaras (13%), Iloilo (6%), and Misamis Oriental (25%). These *viajeros* directly purchase salt from the salt producers for transporting to the market center where they are distributed to other intermediaries or directly sold to institutional buyers. Meanwhile, 6 wholesalers, 13 wholesale-retailers, and 25 retailers were found in Visayas while a wholesaler, 4 wholesaler-retailers and 6 retailers were found in Mindanao. In the Mindanao region, the wholesalers purchase salt from the assembler-wholesalers and sell them to the wholesaler-retailers and retailers for disposal in public markets, convenience stores, or variety stores. With the implementation of the Republic Act No. 8172 (ASIN Law), the assembler-wholesalers perform the iodization of salt before packaging if they will distribute the salt to other provinces.

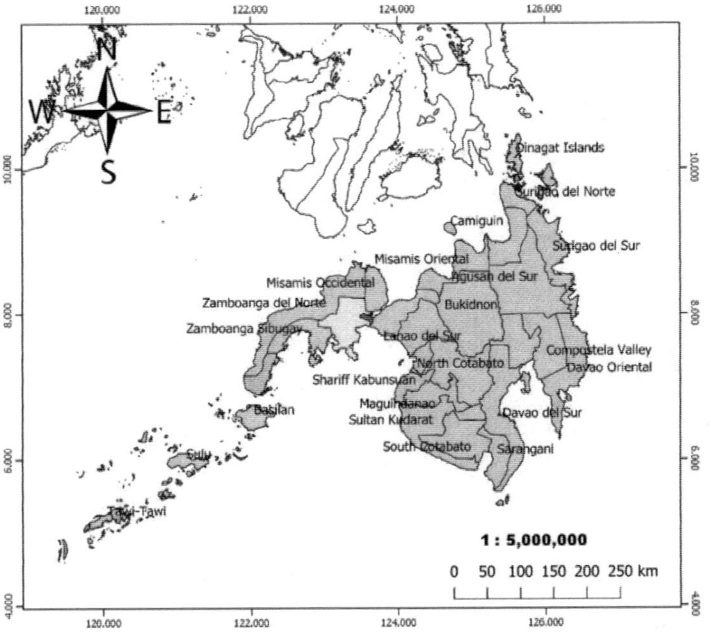

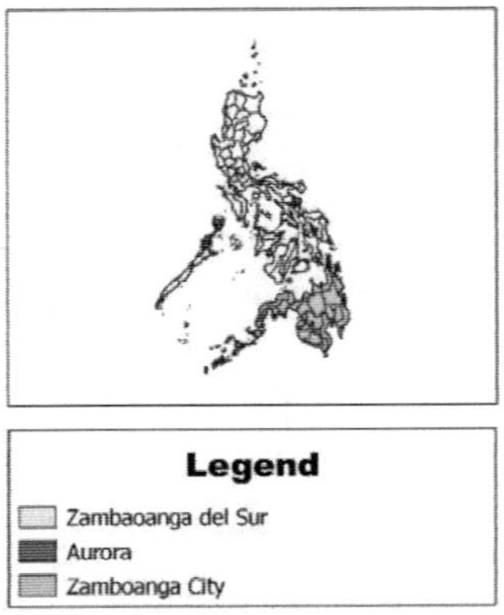

Province	Municipality	Salt Traders
Zamboanga del Sur	Aurora	2
	Zamboanga City	1
	Total	3

Figure 4. Location of market intermediaries in the Mindanao region

The institutional buyers were found in Iloilo (33%) (Figure 5), Negros Occidental (17%) (Figure 6) and Misamis Oriental (50%) (Figure 7). They purchase salt directly from the producers and traders for various applications including the production of fish sauce and fish paste, dried fish, water treatment, and preserving ice. The household consumers, on the other hand, were traced in the demand areas composed of 57 from Iloilo (38%) (Figure 5), 44 from Misamis Oriental (29%) (Figure 8), and 50 from Zamboanga City (44%) (Figure 9).

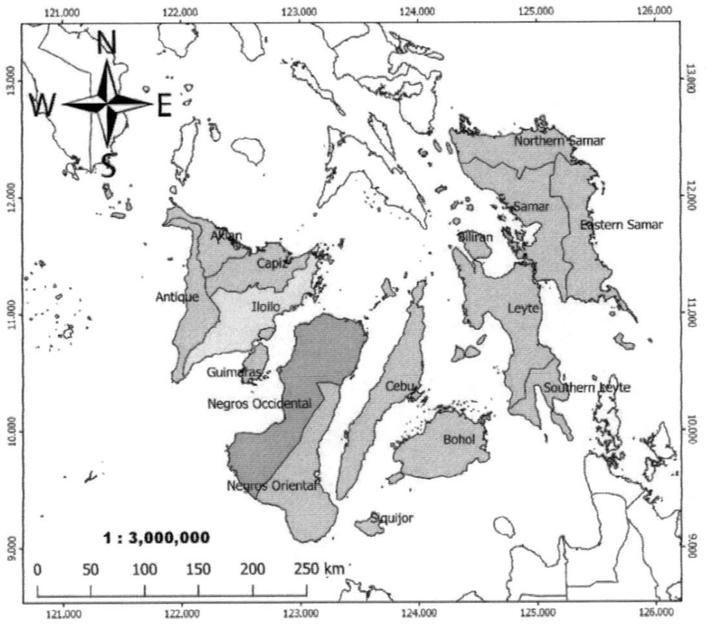

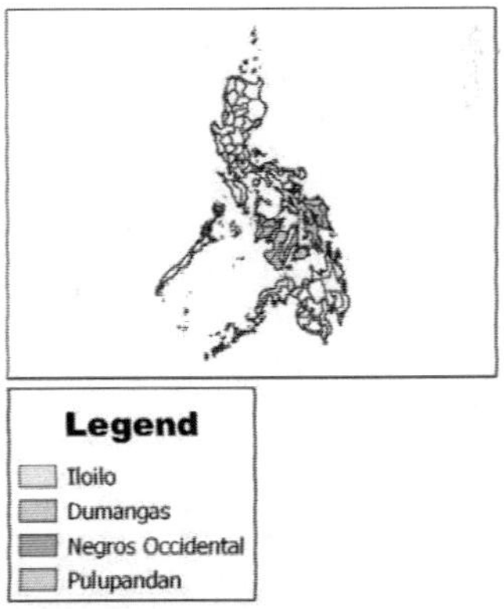

Province	Municipality	Intitutional Buyers
Iloilo	Dumangas	2
	Subtotal	2
Negros Occidental	Pulupandan	1
	Subtotal	1
	Total	3

Figure 5. The geographic location of the institutional buyers in Iloilo City, Iloilo

IOP Conf. Series: Earth and Environmental Science 1006 (2022) 012006 doi:10.1088/1755-1315/1006/1/012006

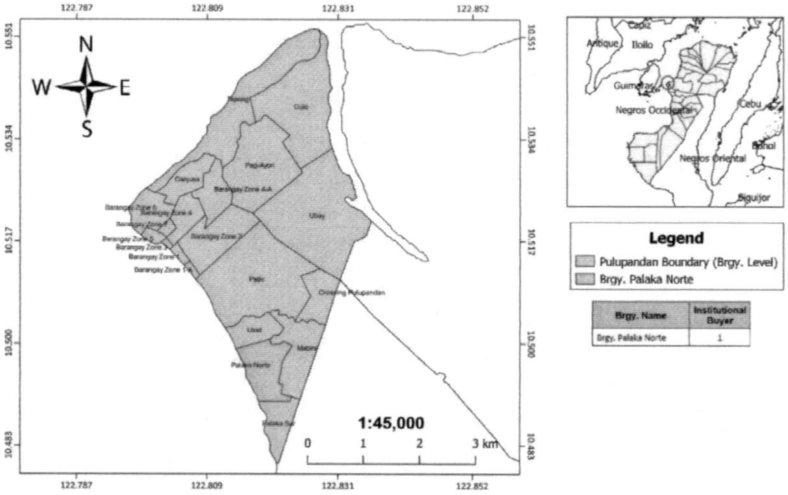

Figure 6. The geographic location of the institutional buyers in Pulupandan, Negros Occidental

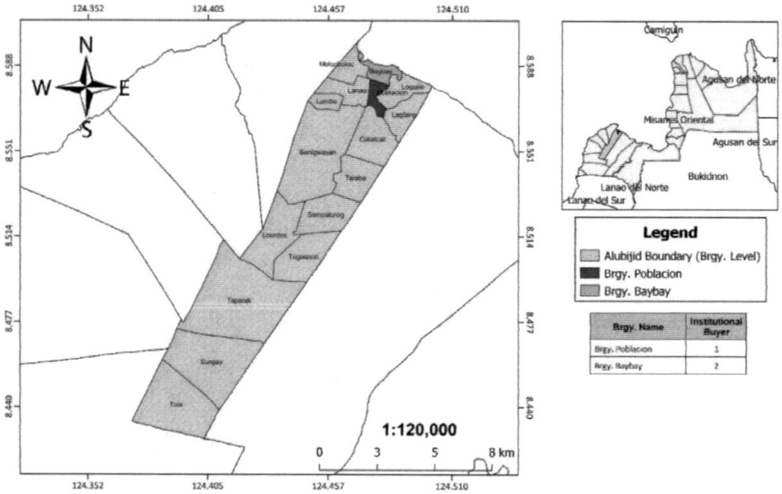

Figure 7. The geographic location of the institutional buyers in Alubijid City, Misamis Oriental

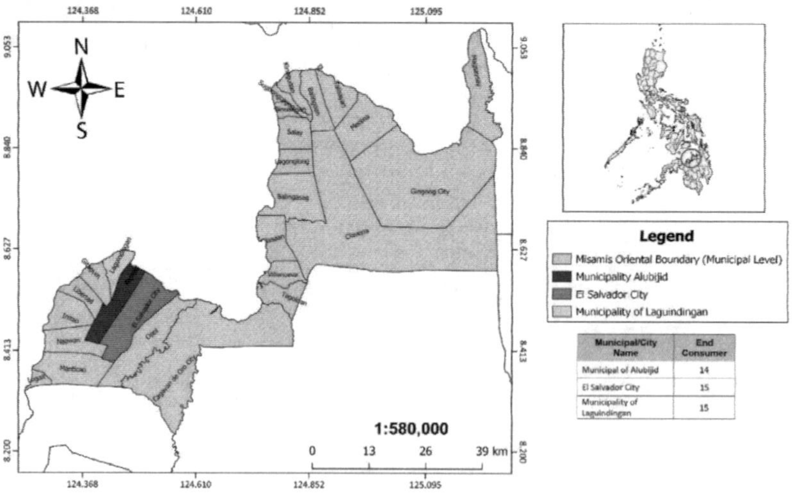

Figure 8. The geographic location of salt consumer in Misamis Oriental

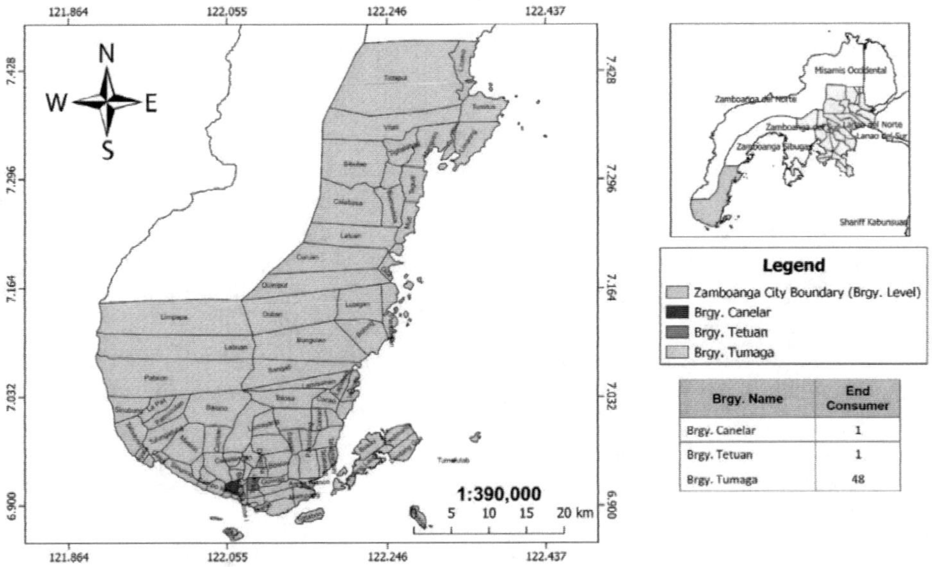

Brgy. Name	End Consumer
Brgy. Canelar	1
Brgy. Tetuan	1
Brgy. Tumaga	48

Figure 9. The geographic location of salt consumer in Zamboanga del Sur

In general, comparing the geographic locations of intermediaries in the two islands, it can be seen in Figures 4 and 5 that they are more scattered and are greater in number in the Visayas (51) than in Mindanao (3). Actual survey done revealed that many small farmers in the Visayas maintain some sort of storage area in their own house where they can keep the harvested salt for some period while waiting to be traded. This means that they are not obliged to sell to only one buyer that comes along and sell the whole lot, they frequently sell to different traders. Also, there are instances when the farmers themselves have their own stalls in the market maintained by their wife which also serve as storage space. In contrast in Mindanao, farmers are tied with a single buyer who comes regularly bringing their own sacks and doing the packing themselves. Salt farmers see this as an advantage in the sense that they will no longer pay for the cost of the sacks and the labor cost for packing/bagging. In addition, in Mindanao, the LGUs are quite strict in implementing salt iodization such that non-iodized salts are not allowed to go far places. Further, comparing Figure 4 (location of market intermediaries in the Mindanao region) and Figure 2 (location of salt farms in the Mindanao region) it can be seen that the market areas (or the salt traders) are concentrated in Zamboanga del Sur particularly in Aurora and Zamboanga City, but the producers are in Alubijid and El Salvador City, Misamis Oriental. It is because there are only three traders found in the area but there were 43 producers traced to have been selling salt to these traders. In addition, as previously mentioned there is limited salt storage facility in the area such that farmers are obliged to sell to the numerous traders plying the area. In contrast, there were 106 producers in the Visayas and only 51 traders. Again, this is because there were farmers whose wife functions as retailers in the public markets.

4. Conclusion and Recommendations

Selected salt farms in Iloilo, Antique, Guimaras, and Negros Occidental in the Visayas and Misamis Oriental and Zamboanga in Mindanao islands were successfully mapped using ArcGIS 10.5 software. Land satellite and google earth images were digitized with the geographic coordinates of the salt farms obtained in situ using a mobile GPS navigator to generate the actual salt field maps. Moreover, the geographic location of the supply chain actors composed of the salt producers, intermediaries, and the end-consumers were also plotted on the map. Their increased visibility on GIS maps will be an effective support tool for determining their population as well as for tracking the geographic flow of salt during production, trading, and consumption. The GIS maps will be useful references for devising salt production suitability maps that will tackle the land and water resources and infrastructure that enables salt farming in the selected study sites. To increase the accuracy of the GIS maps, the exact GPS coordinates of each supply chain actor must also be obtained. Furthermore, developing an open-access salt-mapping system for government entities and other interested entities will greatly help in updating and validating the information on the GIS maps.

5. References

[1] Wang H, Xu X, and Zhu G 2015 Landscape changes and a salt production sustainable approach in the state of salt pan area decreasing on the coast of Tianjin, China *Sustainability* **7** 10078-10097 DOI:10.3390/su70810078

[2] Flad R, Zhu J, Wang C, Chen P, Von Falkenhausen L, Sun Z, and Li S 2005 Archaeological and chemical evidence for early salt production in China *Proceedings of the National Academy of Sciences of the United States of America* ***102*** *35* 12618-12622

[3] Yankowski A 2007 Asinan: Documenting Bohol's traditional method of salt production and the importance of salt in the region's early economy *Philippine Quarterly of Culture and Society* ***35*** *1/2* 24-47

[4] Susanto H, Rokhatia N, and Santos G 2015 Development of traditional salt production process for improving product quantity and quality in Jepara District, Central Java, Indonesia. *Procedia Environmental Sciences* ***23*** 175-178

[5] Nutrition Council of the Philippines (NCP) 2010 A survey of salt importers, producers and traders in the Philippines: An evaluation of internal and external quality assurance and control. Available at http://www.ncp.org.ph/uploads/4/5/5/3/45531383/ncp-2010-salt_survey-final_report.pdf

[6] US Geological Survey in Nationmaster.com 2020 Global trends, statistics, and insights for salt Available at https://www.nationmaster.com/nmx/sector/salt

[7] Verdey A C L and Abilay M J P 2017 DOST eyes salt industry enhancement in occidental mindoro using new technology. Available in the Department of Science and Technology at http://region4b.dost.gov.ph/dost-eyes-salt-industry-enhancement-in-occidental-mindoro-using-new-technology/

[8] Roland A, Erasmus H O and Rosina A K 2019 Impacts of climate variability on salt production in Ghana: Case of Songor salt project *Journal of Sustainable Development* **12**1 1-9

[9] Helmia A and Sasaoka M 2018 Dealing with socioeconomic and climate-related uncertainty in small-scale salt producers in rural Sampang, Indonesia *Journal of Rural Studies* **59** 88-97

[10] Guntur G, Jaziri A A, Prihanto A A, Arisandi D M and Kurniawan A 2018 Development of salt production technology using prism greenhouse method *IOP Conf. Ser.: Earth Environ. Sci* **106** 012082

[11] Bhat A H, Sharma K C and Banday U J 2015 Impact of climatic variability on salt production in Sambhar Lake, a Ramsar, Wetland of Rajasthan, India *Middle East Journal of Scientific Research* **23** 9 2060-2065

[12] Peter B, Messina J, Lin Z, and Snapp S 2020 Crop climate suitability mapping on the cloud: A geovisualization application for sustainable agriculture *Scientific Reports* **10** 15487

[13] Gebeyehu M N 2019 Remote sensing and GIS application in agriculture and natural resource management *Int J Environ Sci Nat Res* **19** 2 45-49

[14] Al-Taani A, Al-husban Y and Farhan I 2020 Land suitability evaluation for agricultural use using GIS and remote sensing techniques: The case study of Ma'an Governorate, Jordan *The Egyptian Journal of Remote Sensing and Space Science* **24** 1 109-117

[15] Refat Faisal B M, Rahman H, Sharifee N H, Sultana N, Imrul Islam M, Ahsan Habib S M and Ahammad T 2020 Integrated application of remote sensing and GIS in crop information system-a case study on aman rice production forecasting using MODIS-NDVI in Bangladesh *AgriEngineering* **2** 17 DOI:10.3390/agriengineering2020017

[16] Sudha Rani D, Venkatesh M N, Naga Satya Sri Ch, Kumar 2018 Remote sensing as pest forecasting model in agriculture *Int. J. Curr. Microbiol. App. Sci,* **7** 3 2680-2689

[17] Yang Y and Zhang S 2008 Approach of developing spatial distribution maps of soil nutrients. In: Li D. (eds) Computer and Computing Technologies in Agriculture vol **1** CCTA 2007 The International Federation for Information Processing 258. Springer, Boston MA

[18] Bandyopadhyay S, Jaiswal R K, Hegde V S and Jayaraman V 2009 Assessment of land suitability potentials for agriculture using a remote sensing and GIS based approach. *International Journal of Remote Sensing* **30** 4 879-895

Acknowledgments
The authors would like to thank the Commission on Higher Education (CHED) for funding this project through the CHED DARETO Research Grant and the administrations of the University of the Philippines and the Cavite State University for the permission to conduct this project.

Evaluating the Effects of Environmental Stress on Leaf Chlorophyll Content as an Index for Tree Health

F Talebzadeh[1] and C Valeo[2]

1 MASc Student, Mechanical Engineering, University of Victoria, BC, Canada
2 Professor, Mechanical Engineering, University of Victoria, BC, Canada
E-mail: talebzadeh@uvic.ca

Abstract. Chlorophyll content plays a vital role in photosynthetic and biomass production in all plants. Because chlorophyll shows a greater sensitivity to changes in external conditions than do other pigments in foliage, such as carotenoids for example, chlorophyll content in leaves may be a good surrogate for environmental stress, changes in temperature and humidity, as well as in pollutant levels both in the air and in the soil. This paper reviews the potential for chlorophyll content in the leaves of trees as a measure of tree health, resistance to stress and environmental conditions. Because chlorophyll content is shown to decrease dramatically with increases in pollution, non-destructive methods for evaluating the amount of chlorophyll in leaves and its changes over a time may be a sufficient indicator for environmental pollutant levels in the air, in the soil and in the water used by a tree.

1.Introduction

Chlorophyll is a natural green compound that is crucial for photosynthetic organisms such as algae, phytoplankton, and all plants. This green, multifunctional substance can be used to predict plant growth, development, above-ground biomass (AGB), and health in various environmental conditions and ecosystems. Chlorophyll is not only important to plant health, but when considering edible plants and fruits, it has a vital role in the human diet [1]. Chlorophyll content which is directly related to above-ground biomass (AGB) is basically controlled and affected by both biotic and abiotic factors in the surrounding environment, such as temperature, soil nutrient, rainfall, sunlight, leaf area index (LAI), environmental pollutions, vegetation type, and human activities [2,3]. Thus, evaluating chlorophyll content and how it changes is important for estimating crops and economic benefits in agricultural ecosystems. But can also be important for signalling environmental stressors such as increases in environmental pollutant levels resulting from anthropogenic activities [4].

Leaf chlorophyll content and changes in content are generally indicators of nutrient deficiencies that may be the result of environmental conditions including pollution concentrations in the canopy [5]. Chlorophyll contents and pigments determine the colour of each leaf, and changes in the amounts of these pigments have been used in various statistical correlations and analyses for inferring plant health, chemistry, and physiology. The literature shows several studies conducted on chlorophyll content suggesting that by monitoring and evaluating chlorophyll content in tree leaves and plants, their responses to environmental stressors, weather and climatic conditions, as well as anthropogenic conditions can be inferred [6]. Chlorophyll content has been shown to be a reliable factor for evaluating plant-life processes and their resistance to stressful conditions such as humidity and temperature changes [7]. One study showed that the decline in chlorophyll content is very fast when plants are in high stress condition and very similar to the changes in pigment during leaf senescence [8]. A decline in chlorophyll content was also reported in a case of water stress due to metabolism increases in plants [9]. Traditional

IOP Conf. Series: Earth and Environmental Science 1006 (2022) 012007 doi:10.1088/1755-1315/1006/1/012007

methods for chlorophyll measurement are done through spectrophotometric techniques or HPLC (High-performance liquid chromatography) measurements. These methods however, require the destruction of leaves and do not permit measurement of chlorophyll changes in a single leave [8]. But advanced technologies that enable easy and fast chlorophyll measurement that are also non-destructive and environmentally friendly have been introduced. For this reason, and the role of chlorophyll in plants health, and chlorophyll content research is becoming more and more prevalent as it can be measured in the field rather easily. Thus, chlorophyll content monitoring is now more than ever a likely candidate for evaluating the effects of environmental stress such as droughts and environmental pollution. The objective of this research is to explore the effects of environmental stressors on chlorophyll contents in trees, while noting the chlorophyll characteristics that are affected by the environment. This would lead to a general assessment of whether chlorophyll content represents a good surrogate for the impact of environmental stress on the plants, and in turn, an indicator of adverse environmental conditions.

2. The effects of environmental stress on chlorophyll content

2.1 Characteristics of chlorophyll and impacts of environmental conditions

Chlorophyll content is one of the most important chemical compounds in plants because photosynthesis and biomass production are controlled directly by the amount of chlorophyll present [6]. Chlorophyll harvests and converts the energy of absorbed photons to the energy of chemical bonds resulting from photosynthesis. Because these pigments are directly related to leaf physiological function, the amount of chlorophyll present in foliage can therefore, provide valuable information on the performance of the leaves toward plant overall health [8]. A chlorophyll molecule is made of pheoporphyrin, which is a magnesium atom that is connected to the nitrogen atoms in the centre of the porphyrin system. there are two different types of chlorophyll: chlorophyll *a* and chlorophyll *b*, and they are both found in all plants and green algae. Chlorophyll *a* has higher amounts of total chlorophyll in plants and thus, the ratio of chlorophyll *a* to *b* is about 3:1 [10]. The amount of chlorophyll in a plant not only determines colour intensity greater chlorophyl content leads to greater biomass production in the plant. Chlorophyll content is also a variety-specific or a species-specific feature, and thus, any changes in chlorophyll content in a plant due to changes in weather or the environmental, are specific to that species and variety of plant.
Studies have shown that chlorophyl content is a reliable indicator of plant response to stressing levels of humidity, precipitation, and temperature. Unfavourable precipitation distribution during the summer growing season months of May to July resulted in a decrease in the concentration of both chlorophyll and carotene pigments [6]. Carotenoids are another group of pigments found in photosynthetic organisms [11] and can be divided in two categories: hydrocarbon carotenes and xanthophylls, which are responsible for creating colours ranging from yellow to red. Although carotenoids are another important pigments in plants, they are masked by chlorophyll and have a minimal influence on plant colour [12]. Carotenoids play a role that protects chlorophyll levels from intensive sunlight. Several studies on carotenoids demonstrated that this part of plants includes vitamin A [6]. Chlorophyll content, however, decreases faster than carotenoid content in instances of environmental stress.

2.2 Air pollution

Chlorophyll contents and pigments are the most likely component of a plant to be affected and damaged by air pollution. These photosynthetic pigments are highly organized and under environmental stress such as air pollution they may undergo photochemical reactions such as reversible bleaching, reduction, and oxidation. For example, chlorophyll reduction in industrial areas is due to degradation of chlorophyll into phaeophytin caused by a lack of magnesium ions [13]. The leaves of many trees have large surface areas that can act as a natural filter to absorb air pollution and remove heavy metal pollution and other airborne contaminants [14]. Research has been done to evaluate the effects of industrial and urban air pollution on the chlorophyll content of leaves. Air pollutants and dust emissions were shown to have a significant impact on the concentration of photosynthetic pigments. Photosynthetic pigments including chlorophyll *a*, chlorophyll *b*, and carotenoids were evaluated by collecting samples from various areas with different levels of air pollution [15]. High polluted areas were located in industrial regions and those with high automobile use. Less or non-polluted areas were chosen in parks, jungles, and Barkatullah University campus in India. A significant reduction in total chlorophyll content was reported in the highly polluted areas with average concentrations observed at approximately 1.04 mg/g, which was less than half of that observed for the unpolluted regions (2.19 mg/g). Also, it was reported that chlorophyll *a* was more affected by air pollution than chlorophyll *b*. The average concentration of chlorophyll *a* was reported as 0.49 mg/g in highly polluted areas. The concentration at the control site was recorded as 1.65 mg/g, which is a reduction of 70.3% in chlorophyll *a*, while chlorophyll *b* showed a 53.78% reduction. The concentrations of heavy metals in atmospheric dust, soil, and leaves were evaluated in 6 different sites in Egypt [16]. Results found a significant correlation between lead concentration in atmospheric dust and leaves in two growing seasons during 2016 and 2017. This study confirmed the main source of lead concentrations in the leaves to be atmospheric dust. This illustrates that air pollutants are readily absorbed by leaves and can easily impact plant health by adversely affecting chlorophyll levels.

2.3 Water and soil pollution

Water and soil pollution are other environmental stresses that can affect chlorophyll content in leaves. Urbanization and industrial growth are known to increase water and soil pollution. Contaminated water applied directly to the plants or contaminated soils are both sources of pollutant in vegetation. Because soil pollutant is received by the plant through the transpiration process and uptake of water, water pollution and soil pollution may be evaluated together in many circumstances involving vegetation health. Heavy metals are one of the most important contaminants present in water and soil pollution [17]. Heavy metals such as nickel, cadmium, zinc, lead, and copper can result in chlorophyll content decrease. In one study [18], it was reported that the lowest chlorophyll content belonged to leaf samples collected from industrial areas. The reduction of 15-66% was directly attributed to Cu, Zn, Pb, Ni, and Cd ion exposure. Metal taken up by plants can concentrate in leaves and in roots [18]. Poplar tree (*Poplar nigra*) leaves are one of the most effective environmental bioindicators due to their sensitivity to heavy metal pollution [19] where heavy metal concentration in leaves are observed to be proportional to the heavy metal concentrations in soil. The relationship between heavy metal pollution in soil and heavy metal concentrations in the leaves of trees was studied in Yazd, Iran [14]. This study confirmed that heavy metals concentrations in leaves can reflect the ability of trees to up take metals from the soil. Cadmium concentrations in tree leaves was 0.34 - 0.62 mg/kg for leaves and was directly related to the soil's metal pollution. While the concentrations of heavy metals in the leaves were different for different trees species, overall, the heavy metals concentration in the leaves resulted from metal uptake from contaminated soil with concentrations in spring and summer being higher than that in the fall - likely due to faster photosynthesis during the spring and summer.

The ability for trees to uptake heavy metals from polluted soil and water can aid in remediating soil and water in contaminated areas [14]. Heavy metals in leaves can potentially damage chlorophyll structure result in adverse effects to the process of natural growth and photosynthesis. Heavy metal exposure can further affect the functional properties of leaves by decreasing mean leaf areas, stomatal pore width, stomatal density, and thickness of the cuticle. All of these parameters are lower in leaves in polluted areas [20]. Although exposure to higher concentrations of contaminants in one study resulted in a reduction in stomatal density from 314 no/mm to 227 no/mm, the shape of the stomatal was normal in both polluted and non-polluted areas [20]. A general trend was reported for heavy metal uptake in leaves and bark in urban trees such as *Ficus nitida* and *Ecualyptus globulus* in Egypt. The concentrations of heavy metals in soil and leaves showed the same trend, and the highest concentration of lead, copper, and cadmium was reported in *Ficus nitida* leaves [16]. Another study in Egypt evaluated the concentration of airborne and soil borne heavy metals on leaves. Results demonstrated that heavy metals concentrations in polluted soil, street dust, bark, and leaves followed similar trends and heavy metal uptake in leaves was observed in the following orders: Zn>Cu>Pb>Cd and Zn>Pb>Cu>Cd, in summer and winter, respectively [21].

3. Conclusions

Stress such as environmental pollution, excessive heat and low moisture levels can have direct and negative impacts on the health of plants. Leaf pigments such as chlorophyll can provide valuable insight into the health and performance of the plant. Research has shown that chlorophyll content can be a useful indicator of plant health because there is an inverse correlation between chlorophyll content and environmental pollutant levels. With regard to heavy metals, which are a common urban pollutant, research shows that plant exposure to heavy metals is possible from the soil, the water and the air. Deposition and absorption are responsible for metal uptake from the air, while transpiration is the driver of metal uptake from water and soil. By using non-destructive devices and methods, chlorophyll content can be an effective way to measure and monitor tree and plant health.

4. References

[1] Q. Jing, B. A. M. Bouman, H. Hengsdijk, H. Van Keulen, and W. Cao, "Exploring options to combine high yields with high nitrogen use efficiencies in irrigated rice in China," *Eur. J. Agron.*, vol. 26, no. 2, pp. 166–177, Feb. 2007, doi: 10.1016/j.eja.2006.09.005.

[2] M. M. Ali, A. Al-Ani, D. Eamus, and D. K. Y. Tan, "Leaf nitrogen determination using non-destructive techniques–A review," *J. Plant Nutr.*, vol. 40, no. 7, pp. 928–953, Apr. 2017, doi: 10.1080/01904167.2016.1143954.

[3] W. Xiong, D. Conway, I. Holman, and E. Lin, "Evaluation of CERES-Wheat simulation of Wheat Production in China," *Agron. J.*, vol. 100, no. 6, pp. 1720–1728, Nov. 2008, doi: 10.2134/agronj2008.0081.

[4] K. M. Keeler *et al.*, "Evaluating the importance of abiotic and biotic drivers on Bythotrephes biomass in Lakes Superior and Michigan," *J. Gt. Lakes Res.*, vol. 41, pp. 150–160, 2015, doi: 10.1016/j.jglr.2015.07.010.

[5] T. Mavromatis, K. J. Boote, J. W. Jones, G. G. Wilkerson, and G. Hoogenboom, "Repeatability of Model Genetic Coefficients Derived from Soybean Performance Trials across Different States," *Crop Sci.*, vol. 42, no. 1, pp. 76–89, Jan. 2002, doi: 10.2135/cropsci2002.7600.

[6] W. Zielewicz, B. Wróbel, and G. Niedbała, "Quantification of Chlorophyll and Carotene Pigments Content in Mountain Melick (Melica nutans L.) in Relation to Edaphic Variables," *Forests*, vol. 11, no. 11, p. 1197, Nov. 2020, doi: 10.3390/f11111197.

[7] L. J. Selzer and C. A. Busso, "Pigments and photosynthesis of understory grasses: Light irradiance and soil moisture effects," *Russ. J. Plant Physiol.*, vol. 63, no. 2, pp. 224–234, Mar. 2016, doi: 10.1134/S1021443716020126.

[8] D. A. Sims and J. A. Gamon, "Relationships between leaf pigment content and spectral reflectance across a wide range of species, leaf structures and developmental stages," *Remote Sens. Environ.*, vol. 81, no. 2–3, pp. 337–354, Aug. 2002, doi: 10.1016/S0034-4257(02)00010-X.

[9] F. A. Begum and N. K. Paul, "Influence of Soil Moisture on Growth, Water Use and Yield of Mustard," *J. Agron. Crop Sci.*, vol. 170, no. 2, pp. 136–141, Mar. 1993, doi: 10.1111/j.1439-037X.1993.tb01067.x.

[10] N. Banu, "Extraction and estimation of chlorophyll from medicinal plants," *Int. J. Sci. Res. IJSR*, vol. 4, pp. 209–212, Nov. 2015.

[11] X. Zhao, T. Jia, and X. Hu, "HCAR Is a Limitation Factor for Chlorophyll Cycle and Chlorophyll b Degradation in Chlorophyll-b-Overproducing Plants," *Biomolecules*, vol. 10, no. 12, p. 1639, Dec. 2020, doi: 10.3390/biom10121639.

[12] R. J. Ritchie, S. Sma-Air, and S. Phongphattarawat, "Using DMSO for chlorophyll spectroscopy," *J. Appl. Phycol.*, vol. 33, no. 4, pp. 2047–2055, Aug. 2021, doi: 10.1007/s10811-021-02438-8.

[13] Govindaraju.M, Ganeshkumar.R.S, Suganthi.P, Muthukumaran.V.R, and Visvanathan.P, "Impact Assessment Of Air Pollution Stress On Plant Species Through Biochemical Estimations," Dec. 2010, doi: 10.5281/ZENODO.1073405.

[14] A. Alahabadi *et al.*, "A comparative study on capability of different tree species in accumulating heavy metals from soil and ambient air," *Chemosphere*, vol. 172, pp. 459–467, Apr. 2017, doi: 10.1016/j.chemosphere.2017.01.045.

[15] S. Giri, D. Shrivastava, K. Deshmukh, and P. Dubey, "Effect of Air Pollution on Chlorophyll Content of Leaves," *Curr. Agric. Res. J.*, vol. 1, no. 2, pp. 93–98, Dec. 2013, doi: 10.12944/CARJ.1.2.04.

[16] A. A. El-Khatib, N. A. Barakat, N. A. Youssef, and N. A. Samir, "Bioaccumulation of heavy metals air pollutants by urban trees," *Int. J. Phytoremediation*, vol. 22, no. 2, pp. 210–222, Jan. 2020, doi: 10.1080/15226514.2019.1652883.
[17] F. Talebzadeh, C. Valeo, and R. Gupta, "Cadmium Water Pollution Associated with Motor Vehicle Brake Parts," *IOP Conf. Ser. Earth Environ. Sci.*, vol. 691, no. 1, p. 012001, Mar. 2021, doi: 10.1088/1755-1315/691/1/012001.
[18] M. Assad, J. Parelle, D. Cazaux, F. Gimbert, M. Chalot, and F. Tatin-Froux, "Mercury uptake into poplar leaves," *Chemosphere*, vol. 146, pp. 1–7, Mar. 2016, doi: 10.1016/j.chemosphere.2015.11.103.
[19] L. Levei *et al.*, "Use of Black Poplar Leaves for the Biomonitoring of Air Pollution in an Urban Agglomeration," *Plants*, vol. 10, no. 3, p. 548, Mar. 2021, doi: 10.3390/plants10030548.
[20] A. Pourkhabbaz, N. Rastin, A. Olbrich, R. Langenfeld-Heyser, and A. Polle, "Influence of Environmental Pollution on Leaf Properties of Urban Plane Trees, Platanus orientalis L.," *Bull. Environ. Contam. Toxicol.*, vol. 85, no. 3, pp. 251–255, Sep. 2010, doi: 10.1007/s00128-010-0047-4.
[21] Y. A. El-Amier and S. Alghanem, "Tree leaves as bioindicator of heavy metal pollution from soil and ambient air in an urban environment," 2018. /paper/Tree-leaves-as-bioindicator-of-heavy-metal-from-and-El-Amier-Alghanem/b00d9636407bb7042e2fc9c49aa6096eefb012e0 (accessed Mar. 24, 2021).

IOP Conf. Series: Earth and Environmental Science 1006 (2022) 012008 doi:10.1088/1755-1315/1006/1/012008

Contribution of Constructed Wetlands for Reclaimed Water Production: A Review

Henrique J O Pinho [1] and Dina M R Mateus[2]

1Smart Cities Research Center (Ci2), Instituto Politécnico de Tomar, Portugal
2Technology, Restoration and Arts Enhancement Center (Techn&Art), BIOTEC, Instituto Politécnico de Tomar, Portugal
E-mail: hpinho@ipt.pt

Abstract. Freshwater scarcity is a growing threat to sustainable development, which can be mitigated by adequate management of water resources. Agriculture and related activities consist in the main use of freshwater, but several other human activities present relevant contributions. Because most of the water uses imply the generation of resultant wastewater, the production and use of reclaimed water by appropriate technologies can be part of the solution to that issue. Considering that the use of constructed wetlands (CWs) can be a relevant contribution to the production of reclaimed water, as an eco-friendly alternative to costly advanced water treatment technologies, this work is a review of the last decade of literature on the use of CWs to produce reclaimed water. The results point to a usual focus on the production of reclaimed water for agriculture or urban spaces irrigation. In order to potentiate a broader application of CWs, some directions of future research and use of this green technology are proposed.

1. Introduction

Freshwater scarcity is a growing threat to sustainable development [1–3]. To cope with this issue, the United Nations identified the objective to ensure water resources management in terms of quantity and quality of water supply and sanitation as the sixth goal of the sustainable development goals [4,5]. Agriculture and related activities consist in the main use of freshwater [6,7], but several other human activities present relevant contributions. Most of the water uses imply the generation of resultant wastewater. Although a significant part of the generated wastewater is not yet subject to any kind of treatment [8], the production of reclaimed water by appropriate technologies can be a relevant contribution to water resources protection and valorisation [9–11].

Reclaimed water consists of wastewater treated by adequate processes that ensure water quality and absence of risk, qualifying the water to be used for some purpose, replacing or decreasing the consumption of freshwater [12,13]. Wastewaters may be generated by different sources, which may be mixed, originating a complex matrix of water pollutants. In general, there are three main types of wastewater: domestic, urban, and industrial [14]. Other minor, but not less relevant types, may also be considered, such as greywater, stormwater, and runoff water. Domestic wastewater is produced in households and similar human urban spaces (Hotels, Restaurants, Public and commercial toilets, and restrooms). It consists of a mixture of blackwater, from toilets, with greywater, from washing and cleaning, bathing, and kitchen. Municipal or urban wastewater is a mixture of domestic with industrial wastewaters. Stormwater includes rainwater, and snow and ice melting, which may cause runoff waters that may include leached compounds originated by the interaction of the water with the soil, landfills, or pavements [15].

The wastewater treatment technologies are diverse, strongly dependent on the wastewater composition and flow rate, and the end-use aim or disposal location. Usually, wastewater treatments are organized in three main stages: primary, secondary, and tertiary or advanced [14]. After the primary stage, and eventual preliminary treatments, the wastewater should be free of large objects and part of the suspended solids, greases, and other organic matter. The aim of the second stage consists of the removal of biodegradable organic matter and may include disinfection. Tertiary, or advanced treatments, aims to remove residual suspended solids and nutrients, and usually include disinfection.
The end uses of reclaimed water may also be diverse and will be described in section 2. Although the quality of reclaimed water for some end uses may be attained with secondary treatments, main reclaimed water end uses require advanced treatments, which may include disinfection [16]. Examples of advanced treatments are additional settling or filtration, coagulation-flocculation, desalination, maturation lagoons, and constructed wetlands (CWs). CWs consist of an eco-efficient and green technology that can contribute to lower the costs of wastewater treatment and potentiate the dissemination of reclaimed water production and reuse [17,18]. Although CWs are usually targeted to treat secondary wastewaters, they can be used to replace secondary treatment and even the primary treatment [19,20]. This option, under a controlled design and operation, can be a solution for decentralized wastewater treatment systems, in which the wastewater is treated and reused closer or at the point of its generation [21], despite the relevance of using CWs in centralized municipal or industrial plants.
There are available some previous reviews on reclaimed water production by CWs. Tao et al. [22] proposed a set of design orientations for using CWs to produce reclaimed water, based on the available data from 31 full-scale CWs. Most of the CWs reported in that work received secondary treated wastewater, river water, or stormwater, while 11 received primary treated wastewater, with the majority producing reclaimed water for non-potable uses. Lavrnić and Mancini [23] analysed the data available in the literature for 29 pilot and full-scale CWs in four Southern European countries and made a comparison of the reported composition of the treated wastewater with the country-existent regulations for reclaimed water reuse in agriculture in 2016. They observed that several treatment plants can provide sufficient water quality for irrigation, but there is a lack of data on some relevant parameters, namely the microbiologic indicators. Almuktar et al. [24] reviewed the CWs characteristics, designs, and efficiencies in wastewater treatment considering the hypothetical reuse of the treated wastewater in agriculture. Shingare et al. [25] made a review on the CWs' capacity to remove pathogens from the wastewater, claiming that is a critical and understudied factor to enable reclaimed water reuse regarding risks of contamination.
The work of Nan et al. [26] focused on the review of published work describing the use of CWs to treat domestic wastewater and consequent evaluation of the potential for water reuse in agriculture. These authors compared the reported quality parameters of the treated water with the current EU criteria to use the reclaimed water for irrigation purposes. They observed that for most CWs systems there is a need for additional treatment steps, and, particularly, the application of disinfection measures. The present work is intended to complement and extend the work of Tao et al. [22], by including published work on studies carried out with pilot-scale CWs in the last decade, and full-scale data not covered by the referred previous review. The data gathered was organized according to the wastewater and CW types, end uses, water quality analysis and experiments carried out to validate the end-use quality criteria.

2. Methods

The review work of Tao et al. [22] that refers to 31 examples of full-scale CWs applied to reclaimed water production was used as starting point. Then, a search in Scopus database was conducted to complete the data set to include pilot-scale and other examples not covered by the referred work. The contents of the extracted documents were evaluated for the focus on the end uses of the reclaimed water produced by CWs. The documents published from 2010 onwards and providing the characteristics of the CWs, wastewater type, analysis performed and water end-use were selected, and then organized according to an adaptation of the end uses of reclaimed water classification following Chen et al. [9] and the main types of wastewater.

3. Results

3.1. Application of CWs by wastewater source and end uses

Figure 1 shows the density of the selected published works over a two-dimensional matrix. One dimension is the end use of the reclaimed water, and the other dimension is the source of the treated wastewater: Domestic, Urban, Industrial, and Others, like greywater and stormwater.

The end uses are organized in: (i) Potable Uses - direct or indirect; (ii), Non-potable Uses - irrigation of animal feeding, food crops, non-food plantations or woodlands, aquaculture production, and urban and residential uses that may include green spaces irrigation, urban lakes, ponds, fountains and recreative spaces, cleaning of urban areas and roads, fire protection, air conditioning, toilet flushing, and washing of cars and clothes, among other urban uses; (iii) Groundwater and Environment - groundwater replenishment, streams recovery and augmentation, seawater intrusion barrier; (iv) and Industrial Uses - such as process water, cooling water, boiler make-up water, and washing of spaces and equipment. Most works focus on the use of reclaimed water for non-potable uses, mainly for irrigation purposes. Table 1 contains references to published work on using reclaimed water for irrigation, which is not covered by the review of Tao and co-workers. The table data includes the type of CW, post-treatment if applied, and the relevant water analysis performed. CWs can be designed in a variety of types, configurations, and combinations [20]. Three main types are common and classified according to the flow arrangement of the water under treatment: subsurface vertical flow CWs, SSVF, abbreviated as VF; subsurface horizontal flow CWs, SSHF, abbreviated as HF; and free surface flow CWs, FSF. When not indicated by the authors, domestic wastewater was considered as urban, and irrigation end-use was classified as for agriculture purposes.

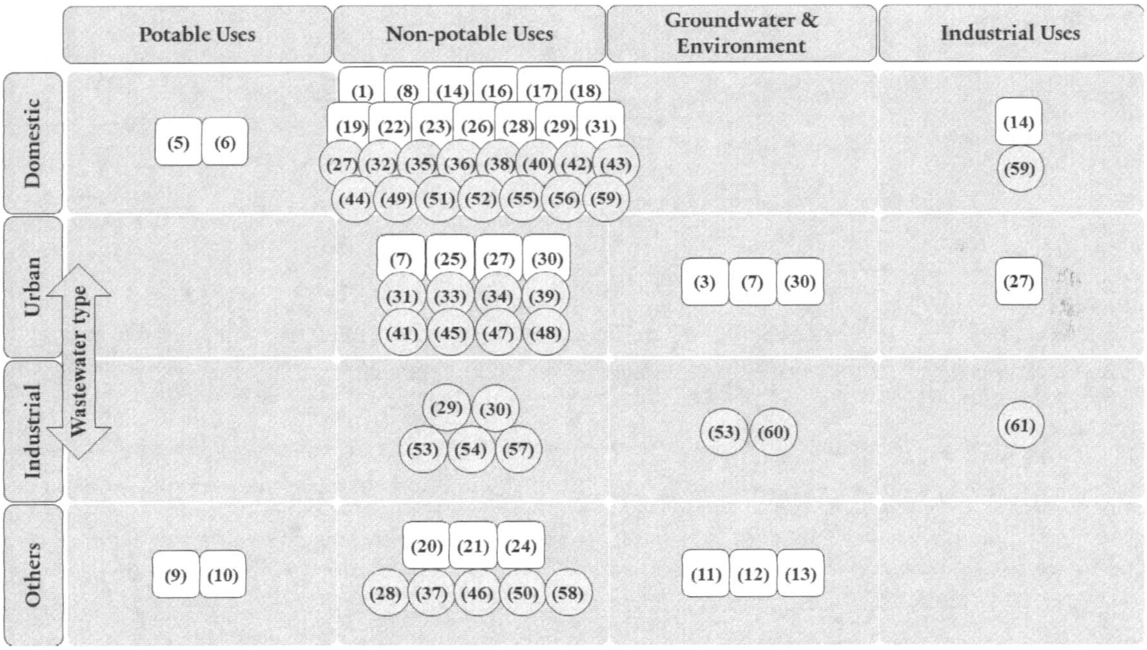

Figure 1. Distribution of the works referred by Tao and co-workers (squares) and the retrieved published works on producing reclaimed water by constructed wetlands in the last decade (circles). The numbers inside squares refer to the works listed in the Tao et al. review [22]. The numbers inside the circles coincide with the reference list of the present work.

From Table 1 it is observed that 70% of the identified works propose the use of the reclaimed water for irrigation in agriculture, but most of this end-use applicability is based on theoretical comparison of the treated water quality with the current legislation (identified with T in the table).

Table 1. Published work on using reclaimed water produced by CWs for irrigation.

Type of wastewater	Type of CW	Post-treatment	Analysis	End-use/test	Ref.
Raw domestic	VF + HF	-	Nut, EC, TSS, SAR, M, O, Mic	Agr/T!	[27]
Artificial raw greywater	VF	-	Nut, TSS, pH, SAR, EC, O, M, Mic	Agr/P!	[28]
Industrial (Pig farm)	VF	-	Mic	Agr/T!	[29]
Industrial (Textile)	HF	-	Nut, TSS, pH, DO, O, M	Agr/P!	[30]
Primary urban	VF	-	Nut, TDS, TSS, O, Mic	Agr/T!	[31]
Secondary domestic	HF	-	Nut, TSS, pH, DO, EC, O, Mic	Urb/T	[32]
Secondary urban	VF + HF + FSF	-	Nut, TDS, TSS, pH, DO, EC, Mic	Agr/T	[33]
Secondary urban	HF	Pond + US + UV	Nut, TSS, O, Mic	Agr/T!	[34]
Secondary domestic	HF	-	Nut, TSS, Mic	Agr/T!	[35]
Primary domestic	HF + VF combinations	-	Nut, TSS, Mic	Urb/T!	[36]
A mix of greywater and wastewater from laboratories	HF + VF combinations	Ponds	Nut, TSS, pH, DO, EC, Mic	Agr/T	[37]
Secondary domestic	HF + VF	-	Nut, TSS, pH, DO, EC, O, Mic	Urb/T!	[38]
Primary urban	HF + VF combinations	-	Nut, TSS, pH, DO, EC, Mic	Agr/T	[39]
Secondary domestic	HF	-	Nut, TSS, pH, DO, EC, SAR, M, O, Mic	Agr/P	[40]
Secondary urban	HF + VF	-	Nut, TSS, pH, Mic	Urb/T!	[41]
Primary domestic	HF	-	Nut, TSS, pH, DO, EC, SAR, M, O, Mic	Agr/T	[42]
Secondary domestic	VF	-	Nut, TSS, pH, DO, EC, Mic	Agr/P!	[43]
Primary domestic	HF	-	Nut, TSS, pH, DO, EC, O, Mic	Agr/T	[44]
Secondary urban	HF	-	Nut, TSS, pH, DO, EC, M, O, Mic	Urb/P	[45]
Raw greywater	HF	-	Nut, TSS, pH, O, Mic	Urb/T!	[46]
Primary urban	VF	Vegetation ponds	Nut, TDS, pH, EC, M, O	Agr/T	[47]
Secondary urban	HF	Filters + Pond or UV	Nut, TSS, pH, EC, SAR, O, Mic	Agr/P	[48]
Raw domestic	HF + VF	-	Nut, TSS, O, Mic	Agr/T!	[49]
A mix of stormwater and primary domestic	HF + FSF	-	Nut, TSS, pH, DO, EC, M, O, Mic	Urb/T	[50]
Secondary domestic	Hybrid + VF	Ozonation	Nut, TDS, pH, EC, O, Mic	Urb/P	[51]
Secondary domestic	HF	Chlorination or UV	Nut, TSS, pH, EC, O, Mic	Urb/P!	[52]
Industrial (Reverse Osmosis concentrate)	HF/VF	-	Nut, TDS, pH, EC	Urb/T	[53]
Industrial (Pig farm)	VF	-	Nut, M	Agr/T	[54]
Primary domestic	VF	Natural UV	Nut, TDS, TSS, pH, DO, EC, O, Mic	Agr/T	[55]
Primary domestic	VF + HF	-	Nut, TSS, pH, EC, Mic	Agr/T	[56]
Industrial (Winery)	VF + HF + FSF	-	Nut, TSS, pH, EC, Mic	Agr/T!	[57]
Greywater	VF	Filter + Chlorination	Nut, TSS, pH, DO, EC, O, Mic	Urb/T!	[58]
Secondary domestic	HF	-	Nut, TSS, pH, DO, EC, M, O, Mic	Urb/P	[59]

Type of CW: HF= Subsurface horizontal flow; VF= Subsurface vertical flow; FSF= Free surface flow; "+"= set of serial or parallel arrangements; "/"= Alternative modes. Type of post-treatment: UV= Ultraviolet; US= Ultrasounds. Analysis: Nut= Nutrients, such as Total Nitrogen, Total Phosphorous, Chemical Oxygen Demand, Biochemical Oxygen Demand, or related analysis; TSS= Total Suspended Solids; TDS= Total Dissolved Solids; SAR= Sodium Absorption Ratio; EC= Electrical Conductivity; DO= Dissolved Oxygen; M= Metals; O= Other analysis; Mic= Microbiologic indicators. End-use/Test: Agr= Irrigation in agriculture; Urb= Irrigation in urban sites; T= Theoretical risk evaluation; P= Practical risk evaluation based on irrigation experiments; "!" Issues identified or abcense of risks not clearly demonstrated.

Only 23% of works attempted evaluating the effects of the reclaimed water by some kind of irrigation experiment (marked as P in the table). Using the theoretical or practical approaches, 48% of the works either identified some risk or had lacking data in the experiments that impeded the risk evaluation (references marked with ! in the end-use test). However, because the most commonly identified risks are related to the microbiological contents of the reclaimed water, disinfection is generally recommended as a solution.

3.2. Cases presenting lower research intensity

As can be seen in Figure 1, the most reported data refers to the use of reclaimed water for non-potable uses (near 77% of total cases), and almost all cases within non-potable uses report to irrigation purposes (more than 95%). Few recent works are focusing on the use of CWs to produce reclaimed water for applications other than non-potable uses in the last decade.

Xing et al. [60] describe a real-scale plant that combines oxidation and stabilizations ponds with free surface flow and subsurface flow CWs to treat papermaking wastewater after advanced treatment by a fluidized-bed Fenton reactor. The reclaimed water is used to restore the water quality of the Shuangji River, in China. In addition to the usual water quality parameters (Nutrients and dissolved oxygen), the dissolved organic matter was characterized, and in-vitro and in-vivo toxicity assays were carried out.

Besides using the reclaimed water for irrigation purposes, Chakraborti and Bays [53] also propose the use of the reclaimed water produced by a pilot-scale combined horizontal and vertical flow CW to feed coastal wetlands. The CW treats wastewater that consists in the concentrate stream of a Reverse Osmosis unit of a water treatment facility. The authors mention an ongoing study of a full-scale three-stage plant consisting of horizontal subsurface CWs, an up-flow vertical flow CW, and a free surface flow CW.

Pinho and Mateus [59] proposed the use of reclaimed water treated by a pilot-scale modular CW for combined irrigation and microalgae production. Although several works focused on the use of algae to treat and recover wastewater, there are no other examples of using reclaimed water produced by CWs. In this work, the microalgae cultivation serves as a complement to reclaimed water valorisation during rainy and cold seasons when the need for irrigation is lower. In addition to demonstrating the absence of toxic effects of the reclaimed water on the microalgae growth, additional toxicity tests were carried out with ornamental plants.

Riggio et al. [61] studied the production of reclaimed water by vertical flow and horizontal pilot-scale CWs. The source of the wastewater is a WWTP treating effluent from an automotive parts industrial plant. The CWs were tested to produce reclaimed water to be returned to the industrial process, diminishing the use of water from wells. Because the electrical conductivity, alkalinity, and calcium content of the reclaimed water did not meet the requirements, the authors concluded that CWs could fit the proposed objectives but require further research on adequate filling media.

4. Perspectives for future research and applications

Based on the potential uses of reclaimed water, and considering the results presented in Figure 1, there are opportunities to explore the CWs' capabilities to contribute to a broader set of applications. Potable and Industrial uses are the less explored in the reviewed literature.

Although the main reported applications of CWs focus on the production of reclaimed water for irrigation purposes, it should be mentioned that the quality criteria attained may be enough for other applications, particularly in the industrial end uses, such as the power sector [62]. The research reviewed, and presented in Table 1, points to the need to improve future work to validate the risks of using reclaimed water.

Some potential uses seem to be less explored, such as the use of reclaimed water for firefighting, and industrial applications. Although the standards for the use in potable applications are stringent, this end-use may represent a relevant market to foster the production of reclaimed water, in which the contribution of CWs can be considered [63]. There are also few reports on the use of CWs to reclaim greywater and stormwater, which can be a potential use of CWs since this technology works well with low-strength wastewaters and is very resilient to non-conventional pollutants [20].

A particular strength of CWs, the multiple modes of design and operation and the consequent adaptability to different wastewaters and reclaimed water applications, is also a barrier to the spread of this technology. As an example of this barrier, all CWs described in this review have different particularities of design and operation. A shorter set of design and operation schemes may foster the spread of the technology, by presenting a set of fewer complex scenarios to the stakeholders. Similarly, the regulations on the production and use of reclaimed water are still diverse, as observed in the reviewed works. The standardization of regulations can also facilitate the dissemination of CWs technology in the water management framework. Moreover, the current efforts to overcome institutional, social, and economic barriers to the use of reclaimed water [64] are an opportunity to be taken into account to leverage the role of CWs.

5. Conclusions

Reclaiming wastewater has a special role in the mitigation of the growing water scarcity threat. Among the available technologies that can contribute to attaining a proper quality of reclaimed water, CWs may have a special role due to their eco-efficiency, low-cost implementation and operation, flexibility, and link to sustainable development and circular economy. Through a review of published work on the production of reclaimed water by CWs in the last decade, combined with a recent review on examples of full-scale reclaiming plants based on CWs, it was observed that most research and fieldwork focus on the use for irrigation or environmental applications, and, mainly, in reclaiming domestic and urban wastewaters. Using other types of wastewater, such as greywater, stormwater, and industrial wastewaters may represent an opportunity to explore CWs implementation. There is still also a need to focus the research on the treated water quality pertaining to these end uses, and a need to standardize reclaimed water regulations and CWs design and operation.

6. References

[1] Vanham D, Hoekstra A, Wada Y, Bouraoui F, de Roo A, Mekonnen M, van de Bund W, Batelaan Q, Pavelic P, Bastiaanssen W et al. 2018 *Sci. Total Environ.* **613–614** 218–32
[2] Yano S, Yamaguchi M, Yokoi E, Kanayama T, Kubota A, Ogawada D, Matsumura A, Gomez-Garcia M, Rutitan V and Oki T 2020 *J. Hydrol. X* **8** 100058
[3] WWAP 2020 *The United Nations world water development report 2020*. Water and Climate Change. UNESCO World Water Assessment Programme. Paris
[4] UN 2015 *Transforming our world: the 2030 Agenda for sustainable development*. United Nations General Assembly on 25 September 2015
[5] WWAP 2018 *Sustainable development goal 6 Synthesis report 2018 on water and sanitation*. UNESCO World Water Assessment Programme. New York
[6] Ritchie H and Roser M. 2017 Water use and stress. Our world in data (https://ourworldindata.org/water-use-stress)
[7] Jones E, van Vliet M, Qadir M and Bierkens M 2021 *Earth Syst. Sci. Data.* **13** 237–54
[8] WWAP 2017 *The United Nations world water development report 2017. Wastewater: The Untapped Resource*. Paris
[9] Chen Z, Ngo H and Guo W 2013 *Crit. Rev. Environ. Sci. Technol.* **43** 1446–516
[10] Fito J and van Hulle S 2021 *Environ. Dev. Sustain.* **23** 2949–72
[11] Lahlou F-Z, Mackey H and Al-Ansari T 2021 *J. Clean. Prod.* **294** 126331.
[12] EU 2020 *Regulation (EU) 2020/741 of the European Parliament and of the Council of 25 May 2020 on minimum requirements for water reuse*. Brussels
[13] da Silva T, Sánchez-Román R, Queluz J and Pletsch T 2020 Treatment options for the direct reuse of reclaimed water in developing countries. In *Wastewater Treatment and Reuse – Lessons Learned in Technological Development and Management Issues*, Ed. P Verlicchi. pp. 1–31 (Elsevier)
[14] Inc. Metcalf&Eddy, Tchobanoglous G, Stensel H, Tsuchihashi R and Burton F 2013 *Wastewater Engineering: Treatment and Resource Recovery* 5th ed., McGraw-Hill Education – Europe
[15] Müller A, Österlund H, Marsalek J and Viklander M 2020 *Sci. Total Environ.* **709** 136125
[16] Salgot M and Folch M 2018 *Curr. Opin. Environ. Sci. Heal.* **2** 64–74
[17] Gorgoglione A and Torretta V 2018 *Sustainability* **10** 3910

[18] Kataki S, Chatterjee S, Vairale M, Dwivedi S and Gupta D 2021 *J. Environ. Manage.* **283** 111986
[19] Verlicchi P and Zambello E 2014 *Sci. Total Environ.* **470–471** 1281–306
[20] Parde D, Patwa A, Shukla A, Vijay R, Killedar D and Kumar R 2021 *Environ. Technol. Innov.* **21** 101261
[21] Jung Y, Narayanan N and Cheng Y-L 2018 *J. Environ. Manage.* **213** 90–7
[22] Tao W, Sauba K, Fattah K and Smith J 2017 *Rev. Environ. Sci. Biotechnol.* **16** 37–57
[23] Lavrnić S and Mancini M 2016 *Water Sci. Tech.* **73** 2616-26
[24] Almuktar S, Abed S and Scholz M 2018 *Environ. Sci. Pollut. Res.* **25** 23595–623
[25] Shingare R, Thawale P, Raghunathan K, Mishra A and Kumar S 2019 *J. Environ. Manage.* **246** 444–61
[26] Nan X, Lavrnić S and Toscano A 2020 *J. Environ. Manage.* **275** 111219
[27] Melián J, Rodríguez A, Araña J, Díaz O and Henríquez J 2010 *Ecol. Eng.* **36** 891–9
[28] Travis M, Wiel-Shafran A, Weisbrod N, Adar E and Gross A 2010 *Sci. Total Environ.* **408** 2501–8
[29] Bonadonna L and Briancesco R 2011 *Int. J. Environ. Health Res.* **21** 331–40
[30] Fibbi D, Doumett S, Lepri L, Checchini L, Gonnelli C, Coppini E and Del Bubba M 2012 *J. Hazard. Mater.* **199–200** 209–16
[31] Abou-Elela S and Hellal M 2012 *Ecol. Eng.* **47** 209–13
[32] Mustafa A 2013 *Int. J. Environ. Sci. Dev.* **4** 20–4
[33] Ávila C, Garfí M and García J 2013 *Ecol. Eng.* **61** 43–9
[34] Toscano A, Hellio C, Marzo A, Milani M, Lebret K, Cirelli G and Langergraber G 2013 *Environ. Technol.* **34** 2327–36
[35] Barbagallo S, Cirelli G, Marzo A, Milani M and Toscano A 2013 *J. Agric. Eng.* **44** 796-802
[36] García J, Paredes D and Cubillos J 2013 *Ecol. Eng.* **58** 57–62
[37] Zurita F and White J 2014 *Water* **6** 213–28
[38] Ayaz S, Aktaş Ö, Akça L and Fındık N 2015 *J. Environ. Manage.* **156** 115–20
[39] Zurita F and Carreón-Álvarez A 2015 *J. Water Health* **13** 446–58
[40] Castorina A, Consoli S, Barbagallo S, Branca F, Farag A, Licciardello F and Cirelli G 2016 *Int. J. Phytoremediation* **18** 626–33
[41] Haghshenas-Adarmanabadi A, Heidarpour M and Tarkesh-Esfahani S 2016 *Water, Air, Soil Pollut.* **227** 28
[42] Andreo-Martínez P, García-Martínez N, Quesada-Medina J and Almela L 2017 *Bioresour. Technol.* **233** 236–46
[43] Almuktar S, Abed S and Scholz M 2017 *Ecol. Eng.* **107** 82–98
[44] Saggaï M, Ainouche A, Nelson M, Cattin F and El Amrani A 2017 *J. Environ. Manage.* **201** 120–8
[45] Licata M, Tuttolomondo T, Leto C, La Bella S and Virga G 2017 *Water Sci. Technol.* **76** 459–70
[46] Ramprasad C, Smith C, Memon F and Philip L 2017 *Ecol. Eng.* **106** 55–65
[47] Ali Z, Mohammad A, Riaz Y, Quraishi U and Malik R 2018 *Int. J. Phytoremediation* **20** 1152–61
[48] Licciardello F, Milani M, Consoli S, Pappalardo N, Barbagallo S and Cirelli G 2018 *Agric. Water Manag.* **210** 232–42
[49] Herrera-Melián J, Borreguero-Fabelo A, Araña J, Peñate-Castellano N, Ortega-Méndez J 2018 *Water* **10** 39
[50] Ventura D, Barbagallo S, Consoli S, Ferrante M, Milani M, Licciardello F and Cirelli G 2019 *Water Sci. Technol.* **79** 1051–9
[51] Colares G, Celente G, da Silva F, de Loreto A, Lutterbeck C, Kist L and Machado Ê 2019 *Water Sci. Technol.* **80** 254–64
[52] Leiva A, Albarrán A, López D and Vidal G 2019 *Water Sci. Technol.* **79** 656–67
[53] Chakraborti R and Bays J 2020 *Water* **12** 158

[54] Dias S, Mucha A, Duarte Crespo R, Rodrigues P and Almeida C 2020 *Int. J. Environ. Res. Public Health* **17** 8592
[55] Santos V, Machado B, Atalla A, Cavalheri P and Magalhães Filho F 2020 *J. Water Health* **18** 1146–53
[56] Torrens A, de la Varga D, Ndiaye A, Folch M and Coly A 2020 *Water* **12** 3139
[57] Milani M, Consoli S, Marzo A, Pino A, Randazzo C, Barbagallo S and Cirelli G 2020 *Water* **12** 1260
[58] Gonçalves R, de Oliveira Vaz L, Peres M and Merlo S 2021 *J. Water Process. Eng.* **39** 101751
[59] Pinho H and Mateus D 2021 *Hydrology* **8** 30
[60] Xing L, Kong M, Xie X, Sun J, Wei D and Li A. 2020 *Sci. Total Environ.* **699** 134369
[61] Riggio V, Ruffino B, Campo G, Comino E, Comoglio C and Zanetti M. 2018 *J. Clean. Prod.* **171** 723–32
[62] Cherchi C, Kesaano M, Badruzzaman M, Schwab K and Jacangelo J 2019 *J. Environ. Manage.* **236** 561–70
[63] Roccaro P 2018 *Curr. Opin. Environ. Sci. Heal.* **2** 46–54
[64] Sgroi M, Vagliasindi F and Roccaro P 2018 *Curr. Opin. Environ. Sci. Heal.* **2** 20–5

Acknowledgments
This work has been funded by national funds through FCT - Fundação para a Ciência e a Tecnologia, I.P., under the Project UIDB/05567/2020.

IOP Conf. Series: Earth and Environmental Science 1006 (2022) 012009 doi:10.1088/1755-1315/1006/1/012009

Multiple Linear Regression Model of Environmental Variables, Predictors of Global Solar Radiation in the Area of East Lima, Peru

Juan J Soria[1], Orlando Poma[2], David A Sumire[2], Joel Hugo Fernandez Rojas[2], Sulamita Marinela Ramos Chipa[2]

1Escuela profesional de ingeniería de sistemas, Universidad Peruana Unión, Lima Perú
2Escuela profesional de ingeniería ambiental, Universidad Peruana Unión, Lima Perú
E-mail: jesussoria@upeu.edu.pe

Abstract. Multiple regression models are very relevant to predict values using predictor variables. The objective of this study was to predict the global solar radiation in the year 2019 in the area of East Lima, Peru. Three continuous quantitative predictor variables were analyzed: temperature, humidity, wind speed and the response variable was global solar radiation, resulting in a model with excellent significance $p<0.001$ that shows the prediction is effective. The multiple linear regression method was used, finding an average global radiation of 175 W/m2 and predictor variables with average temperature of 19.2 °C, humidity 23.9% and wind speed 1.77 m/s, with the highest temperature in summer recorded at 24.6°C, the highest humidity of 51.2% in autumn, the highest wind speed in summer at 2.63 m/s and the highest maximum global solar radiation in spring with 183 W/m^2.

1. Introduction
In the eastern area of Lima there is a meteorological station, on the campus of the Universidad Peruana Unión (UPeU), which allows us to monitor the environmental variables of the sector. The purpose of this study was to evaluate the predictive variables of solar radiation using the multiple regression model. [1] state that "Distribution modeling and mapping was performed to identify key environmental variables and predict the potential range of the species under current climatic conditions in a study conducted in India". They accessed species occurrence data from primary and secondary sources, which they used after cleaning and eliminating bias. Climatic, topographic, cover and soil variables of 1 km resolution were used as environmental predictors.
Predicting a variable from modeling environmental variables by multiple linear regression has been applied in other studies [2], [3] . Modeling is an important step when we want to understand the complexity of environmental behavior and predict changes in the environment. [4] studied the synergy between solar radiation and environmental variables such as temperature. They found that the main factor affecting the deterioration of PCM properties is low temperature. The relationship of solar radiation and temperature was evaluated by [5] in China and [6] in Junín.
The environmental variables in the study are temperature, humidity, wind speed and global solar radiation. Temperature and humidity are very relevant environmental variables in environmental characterization; [7] considered them in a recent study on population, temperature and absolute humidity in settings such as China, England, Germany and Japan. Likewise [8] implemented an intelligent temperature and humidity monitoring system for remote environmental monitoring. [9]

evaluated the accuracy of linear regression equations based on meteorological data such as minimum and maximum temperature, relative humidity, wind speed, solar radiation, extraterrestrial radiation and vapor pressure deficit. In this study we estimate global solar radiation based on environmental variables (temperature, humidity, wind speed) by applying multiple linear regression analysis.

[10] mention in their study on the use of maximum and minimum temperature which estimates solar radiation by equation (1) represented by:

$$\frac{R_s}{R_a} = K_r (T_{\max} - T_{\min})^{0.5} \quad (1)$$

Where Kr is an empirical coefficient that was recommended to be 0.16 for inland regions and 0.19 for coastal regions, associated to a linear regression model described by equation (2).:

$$\frac{R_s}{R_a} = -0.255 + 0.246(T_{\max} - T_{\min})^{0.5} \quad (2)$$

Where Kr is determined locally.

2. Material and methods

This study used a group of 19 500 records of measurements made every 5 minutes, 24 hours a day, in the year 2019, measuring meteorological variables such as global ambient temperature, output humidity and wind speed as predictor variables, and global solar radiation as the dependent one [11] . The instrument used for the measurement of meteorological variable data was the Vantage Pro2 sensor [12]. For this study, the multiple regression method was used to generate the predictive model of global solar radiation suitable for the eastern area of Lima. The analysis was performed by season for the year 2019 (winter, autumn, spring and summer) and efficient results were obtained from the model. [13] mentions the usefulness of the applied method and how it contributed to model problems related to meteorological variables in the area of East Lima, which will be beneficial in exploring the interconnection between the environmental setting, the climate of the earth's atmosphere and what may happen in the future as the climate is monitored, by monitoring meteorological variables.

The predictions of solar radiation have a non-stationary behavior due to multiple parameters and atmospheric conditions that include the large amount and scale of the data involved [14] , which is why this study presents a predictive model based on multiple regression considering meteorological variables that have a great incidence in the global solar radiation is presented, thus considering one of the models that explain the reality in the area of East Lima, Peru.

The normality of the variables under study was tested, using the Kolgomorov Smirnov normality test, as well as homoscedasticity, linearity, independence and non-collinearity, validating the multiple regression assumptions [15]. The relationships between variables were analyzed using simple correlation, yielding a significant relationship between the predictor variables and the response variable [16].

The multiple regression analysis was applied as a means to evaluate statistically significant variables of the measurements made with the Vantage Pro2 instrument during 2019, in which significant descriptive statistics were obtained in the study variables [17].

Figure 1 shows the location of the Vantage Pro2 instrument located at the Universidad Peruana Union campus in Lima Peru that measured the data of the meteorological variables under study during 2019 [18].

IOP Conf. Series: Earth and Environmental Science 1006 (2022) 012009 doi:10.1088/1755-1315/1006/1/012009

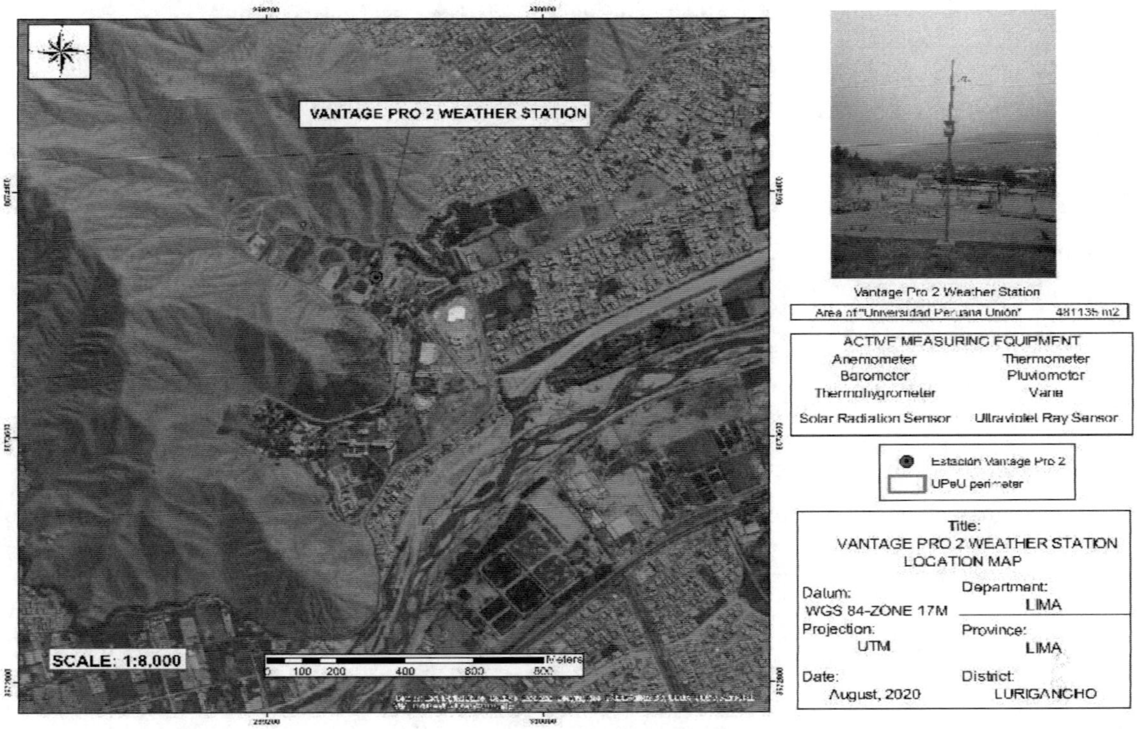

Figure 1. Location of the meteorological station in the East zone, Lima.

3. Results

The descriptive statistics by environmental variable are described in Table 1, which shows an average temperature of 19.2°C, an average humidity of 23.9%, an average wind speed of 1.77 m/s and 175 W/m^2 of global solar radiation, contrasting with the average obtained in the Mantaro Valley with 175.7 W/m^2 studied by [19], there is a mean difference of 0.7 W/m^2.

Table 1. Descriptive statistics of the study variables.

	Temperature(°C)	Humidity (%)	Wind speed (m/s)	Global solar radiation (W/m^2)
N	19551	19532	19564	19564
Missing	13	32	0	0
Mean	19.2	23.9	1.77	175
Median	19.1	20.0	0.00	0.00
Minimun	11.9	0.00	0.00	0
Maximun	29.8	60	16.1	1062

The correlation matrix of the variables is shown in Table 2, which indicates that there is a very good correlation of the predictor variables with the response variable, fulfilling the assumption of the model.

Table 2. Correlation matrix of the study variables.

		Temperature (°C)	Humidity (%)	Wind speed (m/s)	Global solar radiation (W/m^2)
Temperature(°C)	Pearson's r	---			
	p-value				
Humidity (%)	Pearson's r	0.859	---		
	p-value	<0.001			
Wind speed (m/s)	Pearson's r	0.570	0.336	---	
	p-value	<0.001	<0.001		
Global solar radiation (W/m^2)	Pearson's r	0.454	0.182	0.708	---
	p-value	<0.001	<0.001	<0.001	

The environmental variables analyzed recorded an average temperature of 19.2 °C with a minimum of 11.9 °C and a maximum of 29.8 °C; likewise the humidity is presented with an average of 23.9% with a minimum of 0% and a maximum of 60%; also the wind speed with an average of 1.77 m/s with a minimum of 0 m/s and a maximum of 16.1 m/s and solar radiation with an average of 175 W/m^2 with a minimum of 0 and a maximum of 1062 W/m^2. The measurements of the variables by season in the year 2019: in winter an average temperature (15.9 °C), Humidity (6.32%), wind speed (1.28 m/s), solar radiation (164 W/m^2) was recorded; likewise in Autumn an average temperature (22.6 °C), Humidity (51.2%), wind speed (2. 53 m/s), solar radiation (180 W/m^2); also in spring an average temperature (17.3 °C), humidity (9.87%), wind speed (1.27 m/s), solar radiation (183 W/m^2) was recorded and finally in summer an average temperature (24.6 °C), humidity (50.2%), wind speed (2.63 m/s), solar radiation (180 W/m^2) was recorded.

Table 3. Model Coefficients for Global Solar Radiation

			Significance	
Predictor	Estimate	SE	t	p
Intercept	-394.55	10.260	-38.5	< 0.001
Temperature	31.12	0.674	46.2	< 0.001
Humidity	-5.01	0.109	-46.2	< 0.001
Wind speed	52.06	0.577	90.2	< 0.001

In table 3 we observe the coefficients of the multiple regression model with intercept at -394.55 (p<0.001), a temperature coefficient of 31.12 (p<0.001), a humidity coefficient of -5.01 (p<0.001) and a wind speed coefficient with a value of 52.06 (p<0.001), determining the confinable multiple regression model represented in equation (3)

$$y = -394.55 + 31.12x_1 - 5.01x_2 + 52.06x_3 \quad (3)$$

The multiple regression model indicates that for every one unit increase in temperature, global solar radiation will increase by 31.12 W/m^2, likewise for every one unit increase in humidity, solar radiation will decrease by 5.01 W/m^2, and for every one unit increase in wind speed, solar radiation will increase by 52.06 W/m^2.

The 95% confidence intervals of the multiple regression model, the intercept (-394.55) had a variation between -414.66 and -374.44 with a standard deviation of 10.260, the temperature (31.12) a variation between 29.80 and 32. 44 with a standard deviation of 0.674, humidity (-5.01) a variation between -

5.23 and -4.80 with a standard deviation of 0.109 and wind speed (52.06) a variation between 50.93 and 53.19 with a standard deviation of 0.577.

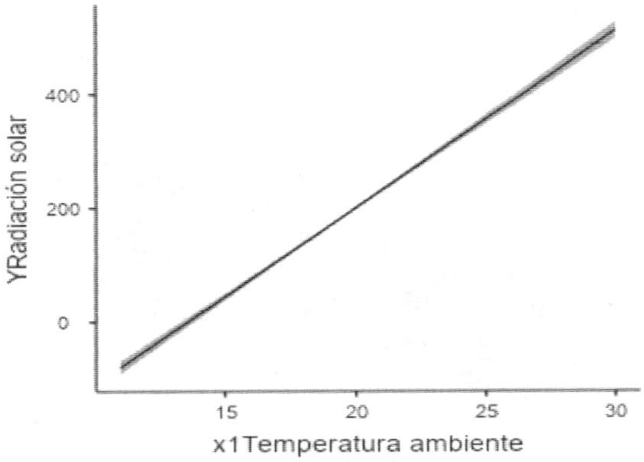

Figure 2. Marginal mean temperature

Figure 2 shows the estimate of the population marginal mean for the temperature predictor variable (19.2 °C), with a 95% confidence interval ranging from 14.8 to 23.6 with a standard deviation of 1.23, which is positively related to global solar radiation.

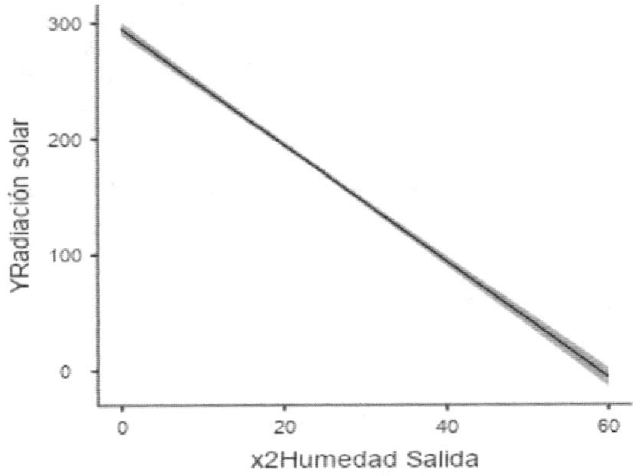

Figure 3. Marginal mean humidity

Figure 3 shows the estimate of the population marginal mean for the humidity predictor variable (23.9242), with a 95% confidence interval ranging from 0.0299 to 47.8186 with a standard deviation of 1.23, which is inversely related to global solar radiation.

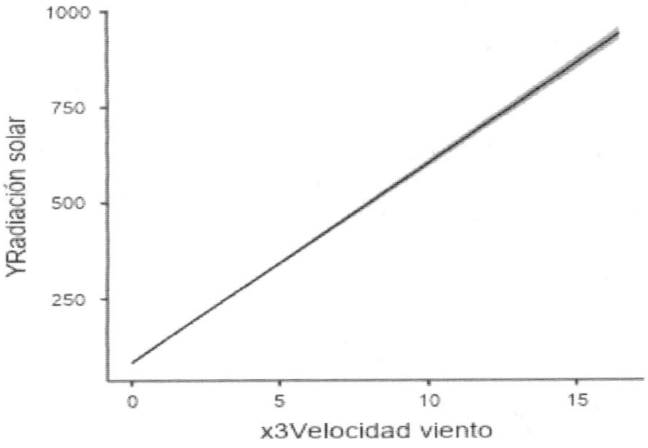

Figure 4. Marginal mean for wind speed

IOP Conf. Series: Earth and Environmental Science 1006 (2022) 012009 doi:10.1088/1755-1315/1006/1/012009

Figure 4 shows the estimate of the population marginal mean for the wind speed predictor variable (1.77), with a 95% confidence interval ranging from -1.03 to 4.57 with a standard deviation of 1.23, which is positively related to global solar radiation.

Figure 5 shows the graph of the residuals of the solar radiation response variable, which shows predictive linearity with a very good level of accuracy (p_value<0.001)..

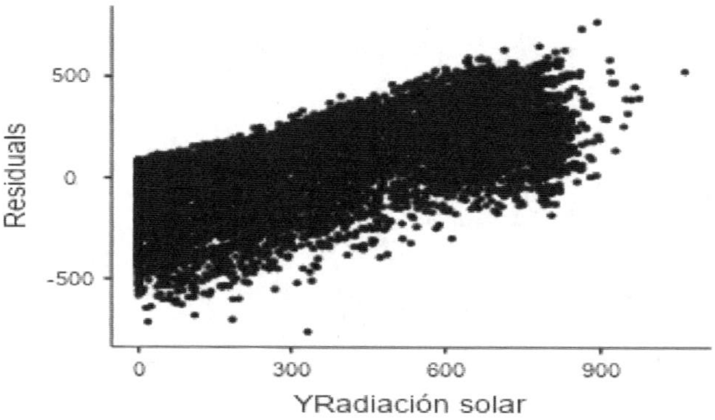

Figure 5. Residuals of solar radiation

Table 4. Model Fit Measures

Model	R	R^2
1	0.746	0.556

In Table 4, we observe that the coefficient of determination R2 of the model explains 55.6% of the original uncertainty and a correlation coefficient of 0.746, which is significant [20].

Table 5. Results of the ANOVA.

Results	Sum of Squares	Mean Square	Significance F	Significance p
Temperature	6.34e+7	6.34e+7	2131	< 0.001
Humidity	6.35e+7	6.35e+7	2132	< 0.001
Wind Speed	2.42e+8	2.42e+8	8142	< 0.001
Residuals	5.81e+8			

In table 5, we observe that all the environmental variables under study have a significance lower than the alpha level α=0.05 (p<0.001) so the multiple regression model is validated at a confidence level of 95%.

Table 6. Sample size and averages by stations

Season	Temperature (Mean)	Humidity (Mean)	Wind speed (Mean)
Winter	6676 (15.9)	6668 (6.32)	6681 (1.28)
Fall	3178 (22.6)	3179 (51.2)	3179 (5.53)
Spring	5564 (17.3)	5551 (9.87)	5570 (1.27)
Summer	4133 (24.6)	4134 (50.2)	4134 (2.63)

Table 6 shows the 19551 records made of the meteorological variables by station in the year 2019 with the Vantage Pro instrument, determining the averages by seaon in its three predictor variables.

4. Discussion

In the study by [10], they found a regression model, based on Hargreaves equation, which estimates solar radiation. Their model is similar with the one found in the present study in its prediction and accuracy in calculating global solar radiation.
The study by [11] mentions that correlation in their regression model generated by the Dumas equation has a correlation coefficient located in the interval $0.8 < r < 0.9$ which shows a good performance of their model. In the environmental predictor variables and global solar radiation, they had a correlation of 0.746 which makes the multivariate linear regression model confinable.
The study by [6] classifies the temperature by seasons of the year, finding that the maximum and minimum temperatures decrease in the months of May, June and July, which applied the Bristow model to estimate the daily, monthly and seasonal solar radiation. Applying the multiple regression model, a temperature of 24.6 °C was found in January, a humidity of 51.2 % in March, a wind speed of 2.63 m/s in January and a global solar radiation of 183 W/m^2 in September 2019.
Regarding the average global solar radiation of the study is 175 W/m^2, in which there is a similarity with the findings found by [19] with a value of 175.7 W/m^2 which makes a mean difference of 0.7 W/m^2.

5. Conclusion

According to the objective of this study, a confinable multivariate regression model ($p<0.001$) was found that allows for the prediction of global solar radiation (175 W/m^2) in the East Lima region based on the environmental predictor variables on average; temperature (19.2°C), humidity (23.9%) and wind speed (1.77 m/s). Seasonal averages were also obtained: the highest temperature was summer with 24.6°C, autumn with the highest humidity of 51.2%, summer with a wind speed greater than 2.63 m/s and spring with a maximum global solar radiation of 183 W/m^2.

6. References

[1] S. Roy, S. Ray, and S. K. Saikia, “Indicator environmental variables in regulating the distribution patterns of small freshwater fish Amblypharyngodon mola in India and Bangladesh,” *Ecol. Indic.*, vol. 120, no. August 2020, p. 106906, 2021, doi: 10.1016/j.ecolind.2020.106906.
[2] K. H. Jung and D. S. J. Kim, “Prediction of pitting corrosion characteristics of al-6xn steel with sensitization and environmental variables using multiple linear regression method,” *Corros. Sci. Technol.*, vol. 19, no. 6, pp. 302–309, 2020, doi: 10.14773/CST.2020.19.6.302.
[3] S. Sharma, R. Gupta, R. Bhatia, A. P. Toor, and H. Setia, “Predicting microbial response to anthropogenic environmental disturbances using artificial neural network and multiple linear regression,” *Int. J. Cogn. Comput. Eng.*, vol. 2, no. March, pp. 65–70, 2021, doi: 10.1016/j.ijcce.2021.03.001.
[4] A. K. Kychkin, O. V. Startsev, M. P. Lebedev, and V. V. Polyakov, “Effect of solar radiation and synergism of the effect of UV radiation, temperature and moisture on the distraction of polymer composite materials in a cold climate,” *Procedia Struct. Integr.*, vol. 30, pp. 71–75, 2020, doi: 10.1016/j.prostr.2020.12.012.
[5] Z. peng XING *et al.*, “Temperature and solar radiation utilization of rice for yield formation with different mechanized planting methods in the lower reaches of the Yangtze River, China,” *J. Integr. Agric.*, vol. 16, no. 9, pp. 1923–1935, 2017, doi: 10.1016/S2095-3119(16)61596-4.
[6] B. F. Camayo-Lapa, D. E. Condezo-Hurtado, A. Y. Ramos-Cadillo, J. R. Massipe-Hernández, and A. B. Camayo-Vivas, “Estimación de la radiación solar global, mediante temperaturas extremas, aplicando el modelo Bristow-Campbell en la región Junín, Perú,” *Ingeniare. Rev. Chil. Ing.*, vol. 27, no. 4, pp. 643–651, 2019, doi: 10.4067/s0718-33052019000400643.
[7] Y. Diao, S. Kodera, D. Anzai, J. Gomez-Tames, E. A. Rashed, and A. Hirata, “Influence of population density, temperature, and absolute humidity on spread and decay durations of

COVID-19: A comparative study of scenarios in China, England, Germany, and Japan," *One Heal.*, vol. 12, no. December 2020, p. 100203, 2021, doi: 10.1016/j.onehlt.2020.100203.

[8] J. Xiao and J. T. Li, "Design and implementation of intelligent temperature and humidity monitoring system based on ZigBee and WiFi," *Procedia Comput. Sci.*, vol. 166, pp. 419–422, 2020, doi: 10.1016/j.procs.2020.02.072.

[9] S. Sharafi and M. M. Ghaleni, "Evaluation of multivariate linear regression for reference evapotranspiration modeling in different climates of Iran," *Theor. Appl. Climatol.*, vol. 143, no. 3–4, pp. 1409–1423, 2021, doi: 10.1007/s00704-020-03473-0.

[10] P. Veeraboina and G. Yesuratnam Guduri, "Analysis of yearly solar radiation by using correlations based on ambient temperature: India," *Sustain. Cities Soc.*, vol. 11, pp. 16–21, Feb. 2014, doi: 10.1016/j.scs.2013.11.004.

[11] A. Dumas *et al.*, "A new correlation between global solar energy radiation and daily temperature variations," *Sol. Energy*, vol. 116, pp. 117–124, Jun. 2015, doi: 10.1016/j.solener.2015.04.002.

[12] Davis Instruments Corp, "USER MANUAL For Vantage Pro2 and Vantage Pro2 Plus Weather Stations," 2020.

[13] J. Cristóbal and S. Alvarado, "UNIVERSIDAD DE CUENCA," Ecuador, 2017.

[14] Y. Zhou, Y. Liu, D. Wang, X. Liu, and Y. Wang, "A review on global solar radiation prediction with machine learning models in a comprehensive perspective," *Energy Conversion and Management*, vol. 235. Elsevier Ltd, p. 113960, May 01, 2021, doi: 10.1016/j.enconman.2021.113960.

[15] D. J. Olive, *Linear regression*. Carbondale, IL, USA ISBN, 2017.

[16] W. K. Härdle and Z. Hlávka, *Multivariate statistics: Exercises and solutions, second edition*. 2015.

[17] S. K. Golfinopoulos and G. B. Arhonditsis, "Multiple regression models: A methodology for evaluating trihalomethane concentrations in drinking water from raw water characteristics," *Chemosphere*, vol. 47, no. 9, pp. 1007–1018, 2002, doi: 10.1016/S0045-6535(02)00058-9.

[18] R. Silhavy, "Advances in Intelligent Systems and Computing 1225," 2020, vol. 2, doi: 10.1007 / 978-3-030-51971-1_33.

[19] W. Clemente De la Cruz and A. Alcántara Boza, "Caracterización de radiación solar directa y difusa en condiciones climáticas del Valle del Mantaro," *Rev. del Inst. Investig. la Fac. Ing. Geológica, Minera, Met. y Geográfica*, vol. 23, no. 46, pp. 87–94, Nov. 2020, doi: 10.15381/iigeo.v23i46.18029.

[20] H. Elorza Pérez Tejada, "Estadística para las ciencias sociales, del comportamiento y de la salud," *Innovación Educ.*, vol. 8, no. 45, 2008.

IOP Conf. Series: Earth and Environmental Science 1006 (2022) 012010 doi:10.1088/1755-1315/1006/1/012010

Possibility for Production of Biogas from Manure in Bulgaria

Ivaylo Ganev[1], Rositsa Velichkova[2], Elitsa Gieva[2], Kalin Filipov[2] and Lubomir Dimitrov[2]

1 Technical University of Sofia, College of Energy and Electronics (CEE), 1000 Sofia, Bulgaria
2 Technical University of Sofia, 1000 Sofia, Bulgaria
E-mail: ivailo.ganev@gmail.com

Abstract. The permanent consumption increasing of energy, foods, techniques and other goods and serviced worldwide leads to the intensifier of their production and to the constant market expansion. In turn, the processes taking place in some of these productions are responsible for the deterioration of the quality of the environment. Over the last decade, science and industry have followed the environmental policies and one of the developing approaches is being the use of by-products as raw materials for new production. The waste to energy technologies meets the requirements both of sustainable managing of wastes and energy production by sources with zero carbon footprint. Typical for this technological approach is its local application and in this study is performed an overview analysis of the possibilities for utilization of livestock manures in systems for production of biogas in Bulgaria. Based on the statistical data for number, and type of farm animals in the country for 2020, their localization, and the potential for gas production of their manure, the calculations were made. The obtained results are presented by regional division and it was estimated that total 1142 installation with 500 kWh capacity could be realized in the investigated farms plants. The obtained results are promising and they are prerequisite for further technological and economic investigations of this waste to energy technology

1. Introduction

The interest in renewable energy sources (RES) increasing because of their inexhaustible and environmental benefits. The biogas fuel is one of the examples for the intensively developed RES and is also part of the concept of circular economy due to the use of resources from agriculture: manure, corn silage and others in Waste to energy technologies. European Union (EU) policy aims to increase production of biogas, biomethane and green hydrogen by the next 30 years to achieve the goal of climate neutrality by 2050. At present, gas fuels obtained as RES provide 7% of gross domestic energy production in the EU [1-3]. Every year, the volume of biogas production in the EU increases by about 20%. At 2007 it reaches a significant growth - 5.9 million tons of oil equivalent. The leading role in this is the extraction of biogas from landfills (49.2%). At the second place is the yield of specially grown energy crops, followed by that in wastewater treatment plants. Typical for the biogas as RES is that it is produced and consumed at the local market which is prerequisite for the different development of this sector European countries [1-3].

Today, Germany occupies a leading position in Europe in the production of biogas. It has more than half of all biogas plants in the EU. According to the German Biogas Association, the number of operating biogas plants in the country is more than 4,000. Most of them have an electrical capacity of 500 kW. Many sites are already in operation, in which the produced biogas is used with the new applications - as fuel for vehicles and for connection to the gas transmission network. At the same time, three of the six largest European companies in the biogas sector are German with a total production volume of about 3.7 million tons. The forecasts for the development of this sector are for intensive increasing in the number of biogas plants in Germany [4-5]. Other leaders in the EU biogas sector are Austria, Denmark, Sweden and the United Kingdom. In Austria, the number of biogas plants increased from about 170 in 2004 to more than 340 in 2005 and almost 600 in 2006. The same trend is established also for the next decade as again most of the installations have a power of 100 to 500 kW. There are official data for 62 landfills for production of landfill gas, 134 installations for sewage sludge processing, 350 installations for biogas and co-fermentation, 25 installations for anaerobic treatment of industrial waste and 15 installations for decomposition of municipal bio waste were in operation in Austria and in the end of last year, a wood biogas plant began operating in the south eastern region. The plant extracts from wood about 100m3 of biogas per hour, enough to heat 150 households. The obtained biogas is not inferior in quality to natural and can be used for heating, electricity generation and automotive fuel. Similar projects are being prepared for implementation also in Germany and Sweden [4-5]. In Denmark is situated the biogas plant which generates heat and electricity by processing residual biomass from more than 75 farms and industrial sites. This plant is characterized with the extracting an energy from a wide range of raw waste materials - fish waste, separately collected household waste, slaughterhouse waste, from the production of beer, alcohol and soft drinks, pharmaceutical waste and contaminated food. The plant generates over 21 million kWh of electricity and over 18.3 million kWh of heat per year, supplying over 1,000 households. The new project for production of grass biogas in Cork, Ireland is also interesting. As the climatic conditions there determine the presence of green grass throughout the year, the municipality decided instead of composting the cut grass, to use it to extract biogas, which is used as fuel for municipal transport. Similar examples can be given from other parts of the world. There are a lot of biogas plants in operation in Japan, China and India. Canada, the United States and South America where the biogas sector is also actively developing. All this undoubtedly proves that biogas technologies are sustainable and economically reliable. With them, every country, municipality, industry, farm, and even household can receive a wide range of benefits - energy, economic, environmental and social [3-5].

In Bulgaria, the production of biogas using biomass and agricultural waste is regulated by legislation, fully synchronized with the European. In 2007 the first power plant for production of electricity from biomass was registered in the town of Lyaskovets, with an installed capacity of 0.285 MW. However, it does not have issued guarantees for the origin of the produced energy for 2007. For the period from 2007 to 2011, 11 biogas plants were built. A sharp jump in the production of energy from biomass was observed in the period 2012–2016 - 29 installations were built, working on the basis of plant and animal waste [6]. Due to the lack of sufficient incentives, the number of facilities for the production of energy from agricultural waste has not increased in recent years. According to the register in the Agency for Sustainable Energy Development, the current biogas power plants are 11. There are near Peshtera, Yambol, Nova Zagora, Calapitsa, Saedinenie, 2 near Pazardzhik, Mezdra, Banya, Parvomay, Plovdiv. Biogas plants prevail in livestock farms [6]. Although there are few installations in our country, unpleasant odors and irregularities in the logistics of biological waste have repeatedly been a cause of civil discontent. The last two challenges have to be overcome for successful implementation of the circular economy conception in the biogas production in Bulgaria with aim to produce energy from farm wastes.

The aim of this paper is to make an analysis based on the statistical data for the possibilities of using manure for biogas production in Bulgaria.

2. Livestock in Bulgaria

Livestock is one of the traditional and most important industries in the Bulgarian economy. In recent years, the pace of development of animal husbandry, according to experts, is clearly defined and measurable - an increase in the number of animals and genetics, which is getting closer to the world levels [7].

In Figures 1-4 are given the number of different livestock: pigs, bovine, sheep and goat in Europe Union for 2020.

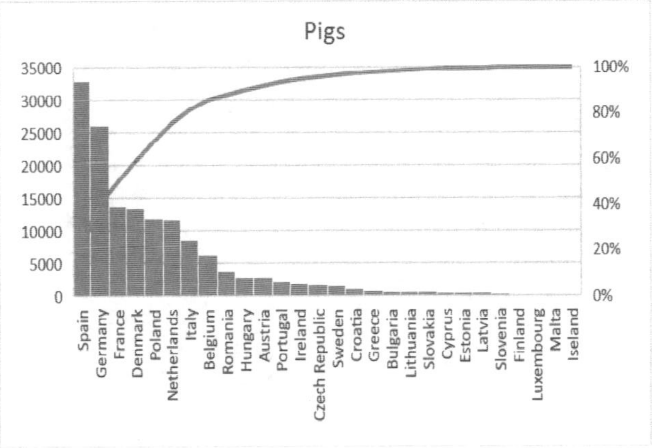

Figure 1. Number of pigs in EU for 2020 in thousand

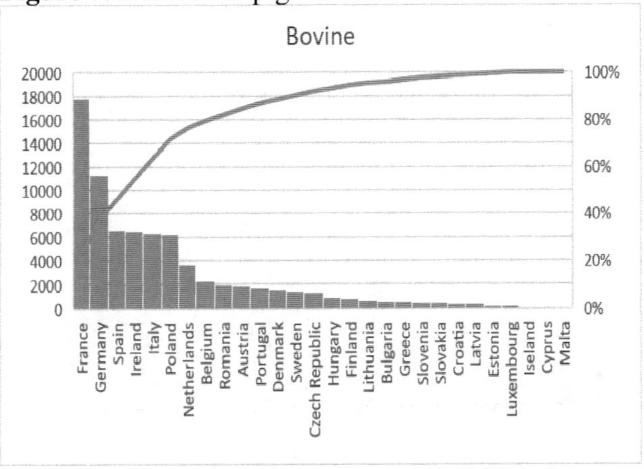

Figure 2. Number of bovine in EU for 2020 in thousand

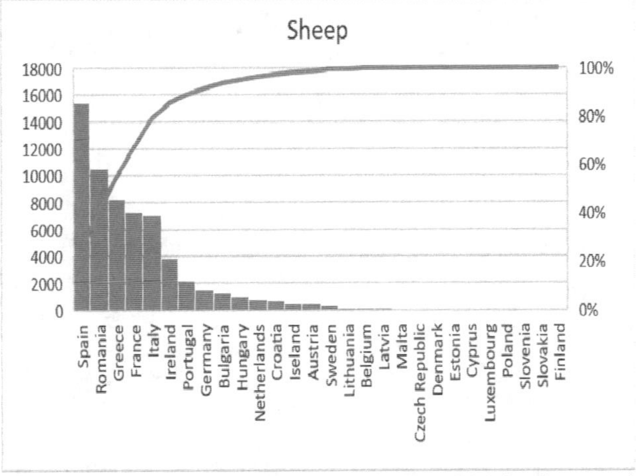

Figure 3. Number of sheep in EU for 2020 in thousand

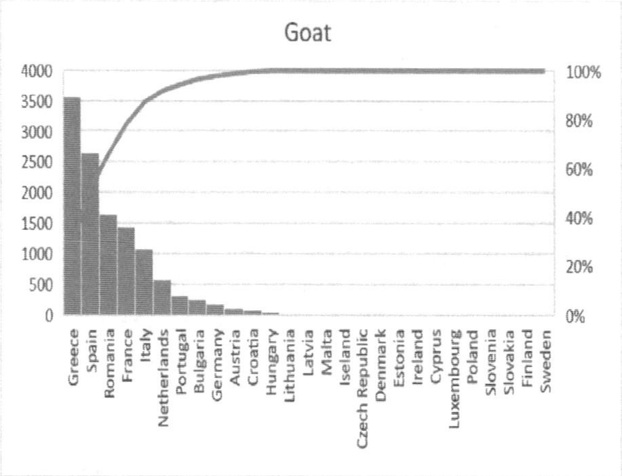

Figure 4. Number of goat in EU for 2020 in thousand

It can be seen that in general the most pigs and sheep are in Spain, bovine in France and respectively the most goats in Greece. From the data for Bulgaria it can be seen that goat and sheep are almost 10 % of their total number in EU. For the other farm animals, the country is placed at the last places. Figures 5-10 represents the dynamic of numbers of livestock and poultry for the period 2015-2020 y in Bulgaria [8-13].

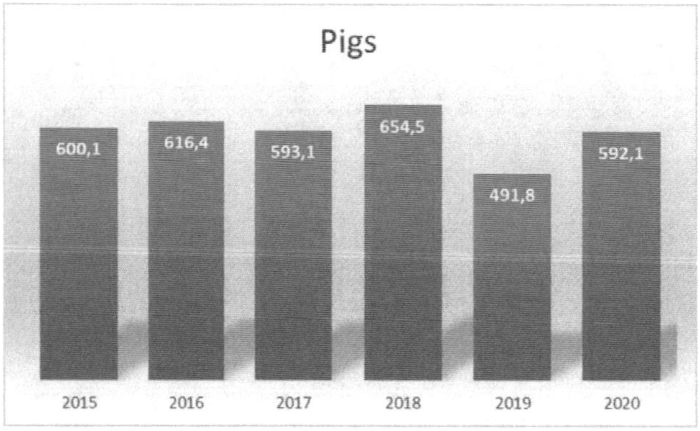

Figure 5. Number of pigs in Bulgaria for period 2015-2020 in thousand

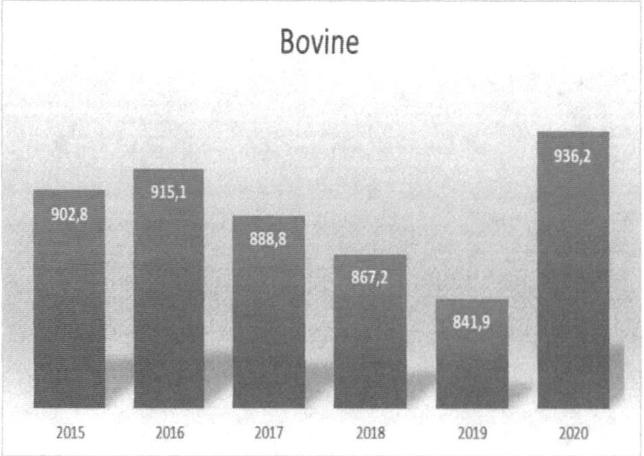

Figure 6. Number of bovine in Bulgaria for period 2015-2020 in thousand

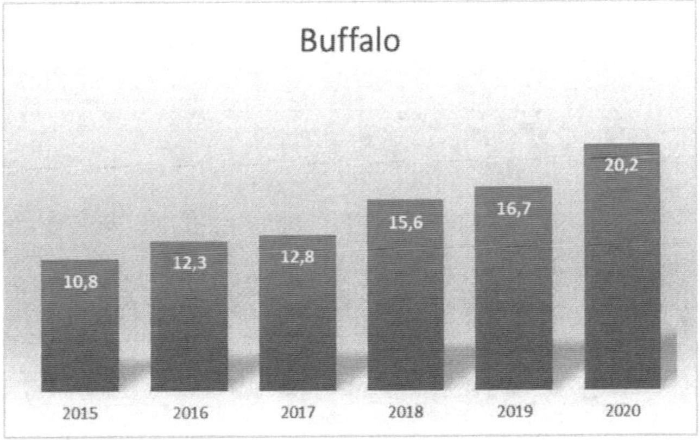

Figure 7. Number of buffalo in Bulgaria for period 2015-2020 in thousand

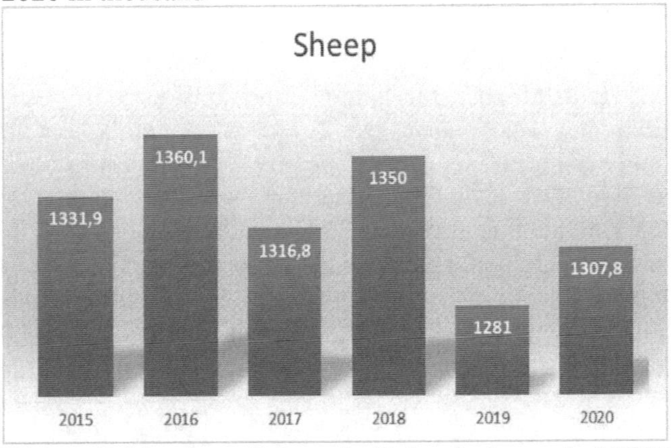

Figure 8. Number of sheep in Bulgaria for period 2015-2020 in thousand

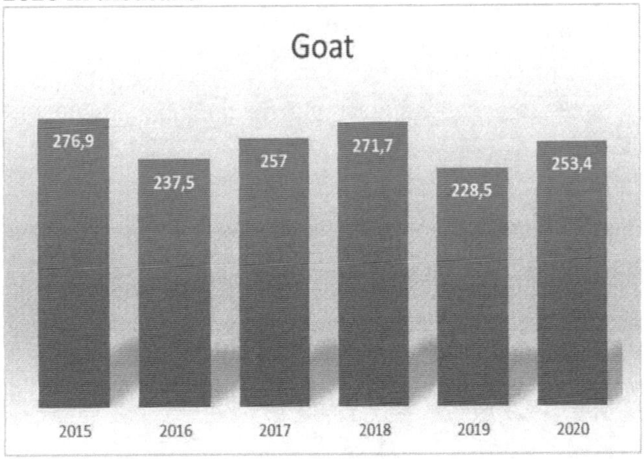

Figure 9. Number of goats in Bulgaria for period 2015-2020 in thousand

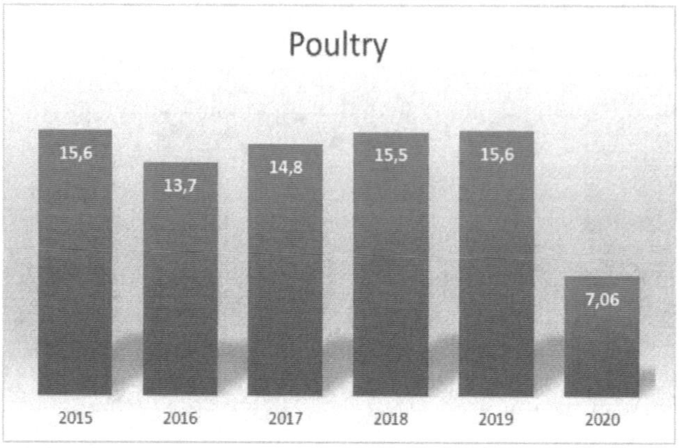

Figure 10. Number of poultry in Bulgaria for period 2015-2020 in millions

The figures show that the most bred livestock in Bulgaria are poultry, pigs and bovine although they have minor share in the EU farm animals. During the period considered it is established a movement in the number of bred livestock, but in general the trend is for permanently kept animals. In 2020 there is a certain increase in the number of bred livestock. The only exceptions are poultry, which have almost double decreasing compared to other years. The deviation during the 2020 could be associated also with the drastically changed food market due to the Covid-19 pandemic.

It is analysed also the geographical data of farm animals and in Figures 11-15 is given the distribution of livestock and poultry according to the nomenclature of territorial units for statistics (NUTS) 2 in Bulgaria [8-13]. NUTS 2 basic regions for the application of regional policies in Bulgaria are defined as: 1-North Western region, 2-North central, 3- North Eastern, 4-South Eastern, 5-South Central, 6-South Western.

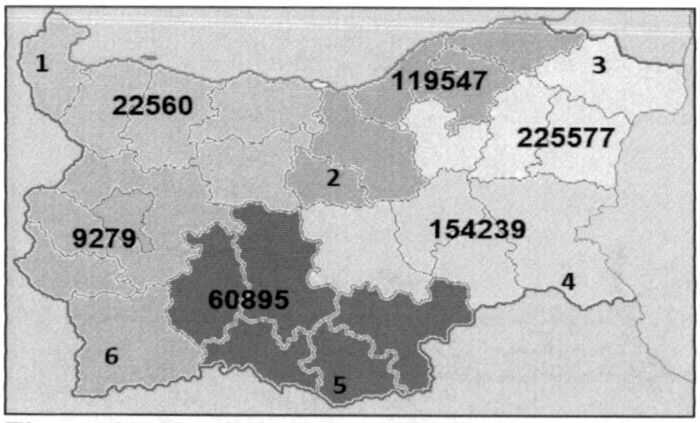

Figure 11. Number of pigs in Bulgaria for 2020 at different NUTS 2 region

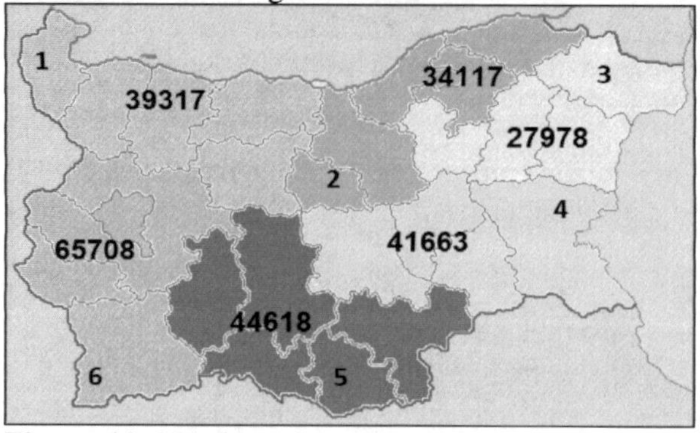

Figure 12. Number of goats in Bulgaria for 2020 at different NUTS 2 region

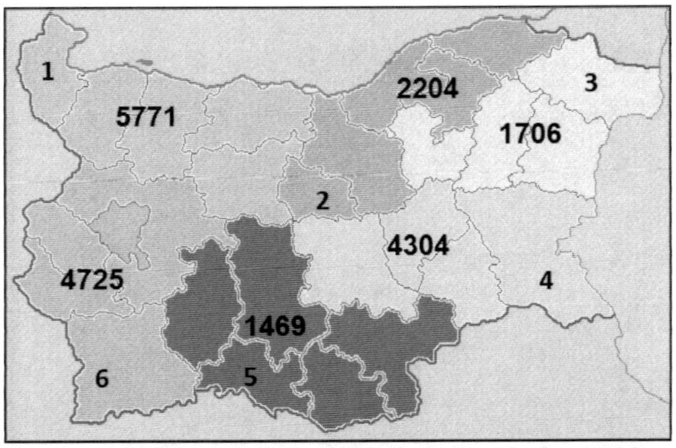

Figure 13. Number of buffalo in Bulgaria for 2020 at different NUTS 2 region

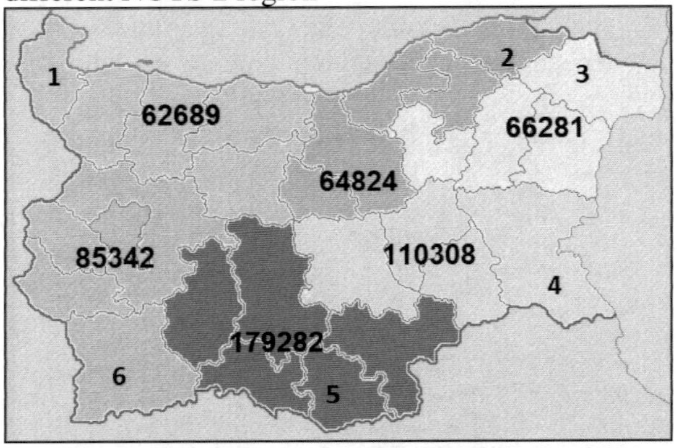

Figure 14. Number of bovine in Bulgaria for 2020 at different NUTS 2 region

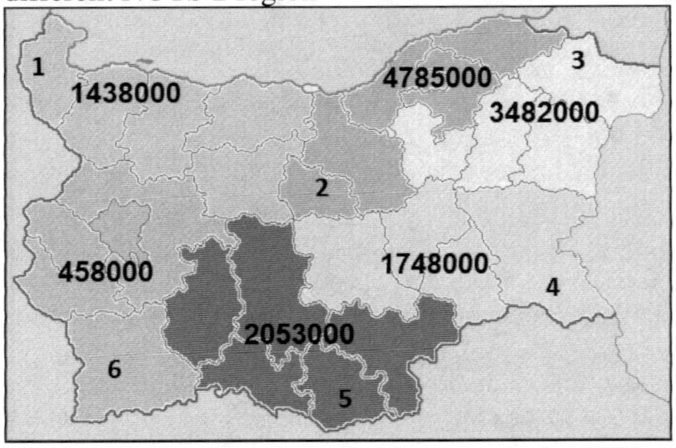

Figure 15. Number of poultry in Bulgaria for 2020 at different NUTS 2 region

3. Analysis of the possibilities of using manure for biogas production in Bulgaria

A forecast calculation of the quantities of waste from the most massive farm animals has been made in this study on the basis of daily waste from the respective animal: from cattle - 35-40 kg of manure; from a pig - 4.5 kg; from goat or sheep - 1 kg; from 1 hen - 0.17 kg. The obtained annual weights of solid waste produced from one animal are: cow ≈ 12.76 t/year, sheep or goat ≈ 0.365 t/year, pig ≈ 1.6 t/year, chicken / hen ≈ 0.06 t/year.

Taking into account the approximate annual values of waste per animal and also the number of animals bred by species, we have calculated the total volume of waste generated by livestock in the country for 2020 and the results are listed in table 1.

Table 1. Total amount of manure for 2020 in Bulgaria

Livestock	*Amount manure, tons*
Bovine	*18724000*
Pigs	*947360*
Sheep	*477347*
Goats	*92491*
Poultry	*423600*
Total	***20664798***

Based on the assumption that 12,000 tons of input material per year are needed to supply one installation with a capacity of 500 kWh, the obtained results for waste of 20 million tons per year for 2020 show that there are raw material sufficient to supply and maintain in operation about 1142 installations with a capacity of 500 kWh.
Figure 16 shows the installations that can be built according to forecasts and by regions in Bulgaria.

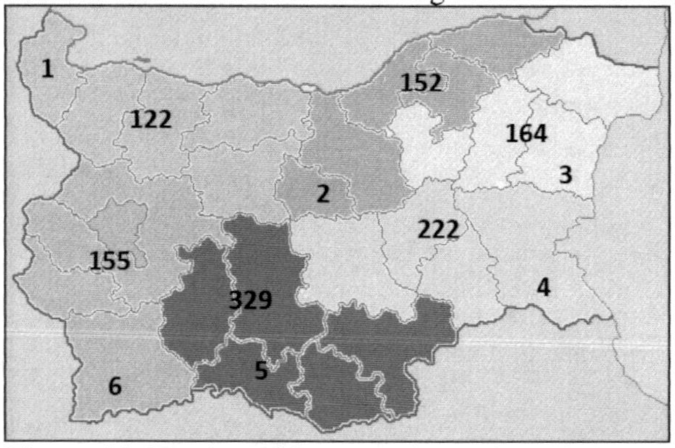

Figure 16. Number of forecast bio gas installation by regions

Another forecast study of the quantities of biogas that can be obtained from animal manure for 2020 in Bulgaria has been made. Regarding to the biogas, the term "animal unit" has been introduced. One animal unit provides manure for 24 hours, which is a raw material capable of producing 1.5 m^3 of biogas. This quantity is equivalent to those obtained by 1 cow, 6 pigs and 250 hens. Based on the data for 2020, the estimated amount of biogas that can be produced in the regions in Bulgaria has been calculated and presented in Figure 17.

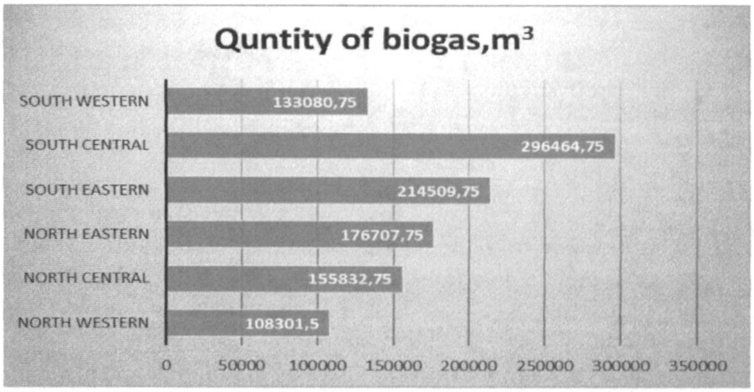

Figure 17. Estimated quantities of biogas which could be obtained by manure calculated for different regions in Bulgaria for 2020

4. Conclusion

The utilization of farm wastes in the context of the circular economy have to be considered at local level to meet the requires for technical and economic efficiencies. In this study the possibilities of using manure for biogas production in Bulgaria are presented based on the statistical overview of number of different animals. Based on the individual waste production by animal type and the manure energy equivalent respectively, the potential number of biogas plants with 500 kWh capacity was established. The obtained data is presented for different regions in Bulgaria according to the distribution of livestock. The results are promising for the development of waste to energy technology through manure utilization which could improve both economic and energy efficiency of farms. Further detailed studies will be performed for the determination of optimal technological conditions for production of synthetic biogas in different Bulgarian farms. For the energy utilization of obtained gas products will be analyzed and economic parameters of the local market.

5. References

[1] Directive 2018/851 of the European Parliament and the Council on amending Directive 2008/98/EC on waste
[2] Directive 2018/850 of the European Parliament and of the Council amending Directive 1999/31/EC on the landfill of waste
[3] http://www.afdc.energy.gov/data/10332
[4] Z. Lajdova, J. Lajda. P. Bielik, (2016). The impact of the biogas industry on agricultural sector in Germany. Agricultural Economics–Czech, 62, 1-8.76(2016)
[5] Analysing Bioenergyand Land Use Competitionin a Coupled Modelling System: The Role of Bioenergyin Renewable Energy Policyin Germanyby Ruth Delzeit, HorstGömann, KarinHolm-Müller, PeterKreins, Betti naKretschmer, JuliaMünch, SonjaPetersonhttps://www.ifw-members.ifw-kiel.de/publications/analysing-bioen ergy-and-land-use-competition-in-a-coupled-modelling system-the-role-of-bioenergy-in-renewable-energy-poli cy-in-germany/kwp-1653.pdf
[6] https://www.moew.government.bg/bg/nacionalen-plan-za-upravlenie-na-otpaducite-npuo-v-republika-bulgariya-za-perioda-2021-2028-g
[7] https://ec.europa.eu/eurostat/web/products-eurostat-news/-/ddn-20200923-1
[8] Agricultural handbook, Ministry of Agriculture and Foods of the Republic of Bulgaria, 2015
[9] Agricultural handbook, Ministry of Agriculture and Foods of the Republic of Bulgaria, 2016
[10] Agricultural handbook, Ministry of Agriculture and Foods of the Republic of Bulgaria, 2017
[11] Agricultural handbook, Ministry of Agriculture and Foods of the Republic of Bulgaria, 2018
[12] Agricultural handbook, Ministry of Agriculture and Foods of the Republic of Bulgaria, 2019
[13] Agricultural handbook, Ministry of Agriculture and Foods of the Republic of Bulgaria, 2020

Acknowledgments

The research leading to these results has received funding from the Ministry of education and science under the National science program INTELLIGENT ANIMAL HUSBANDRY, grant agreement n°Д01-62/18.03.2021

IOP Conf. Series: Earth and Environmental Science 1006 (2022) 012011 doi:10.1088/1755-1315/1006/1/012011

Photovoltaic Panels and Solar Collectors in Cityscape of Bialystok (Poland) and Cordoba (Spain)

Dorota Gawryluk[1], Dorota Anna Krawczyk[2] and Antonio Rodero [3]

1 Bialystok University of Technology, Faculty of Civil and Environmetal Enginnering, Institue of Architecture and Town Planning, Bialystok, Poland
2 Bialystok University of Technology, Faculty of Civil and Environmetal Enginnering, Department of HVAC Engineering, Bialystok, Poland
3 School of Engineering Sciences of Belmez, University of Cordoba, Cordoba, Spain
E-mail: d.gawryluk@pb.edu.pl; d.krawczyk@pb.edu.pl; a.rodero@uco.es

Abstract. The technologies using of renewable sources energy like photovoltaic panels and solar collectors were used in Europe, mainly in southern countries (like Spain) until recently. EU policy of sustainable development and technological development impact on increase of popularity of those technologies in northern European countries like Poland. The research problem of the work is to determinate the appropriate conditions for the location of above-mentioned technology (panels) in the cityscape. The numerous examples of its destruction were noticed because panels were located using only energy efficiency conditions. The research methodology was based on "in situ" research (Bialystok-Poland, Cordoba – Spain) in order to determinate the categories of urban landscape in which panels are located. Technical, climatic, landscape and composition conditions were analysed in order to establish assessment of completed investments. The result of the science work are categories of the urban landscape views in Bialystok and Cordoba, ie architectural and landscape encloses (streets, squares, public gardens), city panoramas. The proposed methodology of assessment of cityscape impact are applicable to the planning, designing and investment process for Bialystok and Cordoba. They could be used to another cities also in necessary condition of taking into account the individual characteristic of cityscape. The proposed conclusions can be used in the planning, designing and investment process of another cities due to protection of their cityscape values.

1. Introduction

Recently, technologies of photovoltaic (PV) panels and solar collectors were used mainly in south Europe, including countries with a high solar radiation, e.g. in Spain. Nowadays, these technologies are used more and more often in northern European countries, including Poland. This is the result of technological development and the EU's policy on sustainable development, including the use of renewable energy sources.

Solar collectors and PV panels are installed on roofs, ground, walls or as a part of facades, depending on technical possibilities, architectural and local conditions. All solutions have advantages and disadvantages. Velasco et al. [1] studied different oriented facades in Spain (Barcelona) to find reductions in radiative heat gains entering the building and the heat recovery in the Venetian blind collectors. Bandaru et al. [2] conducted a deep review of different PVT systems for residential applications and described their performance indicators, progress, limitations and research opportunities. Topic of integrating solar photovoltaics in residential buildings into zero energy buildings in Hail City,was explored by Abdelhafez et al.[3]. Visa et al. [4] studied and commented architectural acceptance of façade integrated solar-thermal solutions for the flat plate solar thermal collectors, in terms of shape, size, color. Coloured BIPV technologies were researches also by Pelle et al [5]. The effective use of photovoltaic panels in order to obtain the appropriate aestetic effect in architecture was considered in the research by Celadyn and Filipek [6]. Harmonised and disharmonised solutions were analised there. A methodology for determining the solar energy potential of large scale urban areas based on building typology was presented in [7]. In turn in paper [8] shows the possibility of solar energy utilization on the relationship between urban morphology and solar radiation gain on building rooftops and wall surfaces. Takebayashi et al. concluded that in the city center and in the residential areas, solar radiation on their roofs was reduced to the effect of shadows by surrounding buildings. Moreover authors wrote that in the narrow north– south streets, solar radiation on the walls was reduced when the opposite building was higher than the building in and functionality. Aimilios et al. [9] studied possibilities of architectural integration of active solar systems on the facades and roofs of existing buildings. They analyzed issues of architectural design and integration. They concluded that the architectural integration of PV panels in building facades presented significant economic and ecologic advantages, however the main difficulty of the integration of active solar systems on building facades derived from technical limitations like overshadowing. Contribution, potential and future challenges of building-integrated photovoltaics (BIPV) to the concept of nearly zero-energy cities in Europe were analised by Gholami, Rostvik and Steemers [10]. Tsoumanis et al. [11] studied the smart evolution of historical cities in context of using photovoltaics while respecting cultural heritage. Rosa [12] analised issue of building-integrated photovoltaics (BIPV) in historical buildings especially in context of implementation of somewhat controversial photovoltaic systems on heritage monuments. Deployment view of the neighboring photovoltaic system was topic of research work of Radovanovic and Popovic [13]. Authors considered its degrading effects on operation of neighboring photovoltaic systems in urban environment. Issue of integration of photovoltaics in buildings addressing technical and formal aspects was considered by Schultze in 2013 already [14]. The author, among other things, drew attention to the essemce of the aestetic meaning of placing panels on buildings (walls, roofs or around the building) Photovoltaics on landmark buildings, especially important to cityscape was analysed by Devetakovi´c et al. [15] in context of its distinctive geometries. Research team studied architectural integration quality, context sensitivity and system visibility of the landmark buildings.

There are more and more investments in placing photovoltaic panels and solar collectors on buildings (on roofs and walls) or as an integrated part of facades. This development of solar technologies in urban areas is supported by investment refunds from many EU programms. The amount and shape of installations depend on local conditions, energy needs and results of a system optimization to achieve the best efficiency. Anyway, there is a need to determine the location conditions of the elements of the installation, which aim at protecting the values of the city landscape.

The aim of this paper is to compare technical, climatic, landscape and composition conditions for solar collectors and PV panels installed in Bialystok and Cordoba - in different climatic zones: Poland and Spain.

2. Methodology

The research problem of the work is the assessment of the location conditions of the above-mentioned technology in the city landscape, due to numerous examples of its destruction. The research methodology was based on in situ research (Bialystok, Cordoba) in order to determine the category of urban landscapes in which the elements of the above-mentioned installations are located. Technical, climatic, landscape and compositional conditions were analysed in order to establish interdisciplinary models for design guidelines for the location of panels.

The types of location (architectural and landscape interiors) of photovoltaic panels in the cityscape were determined on the basis of the results of in situ surveys carried out in Bialystok and Cordoba in 2018-2021. Technical and climate conditions were considered as fulfilled in the analysed examples. The study was devoted to cityscape and composition conditions, which were considered as not analysed by many installation designers yet.

The criteria were understood as:
1) landscape (regarding the visibility of panels in cityscape in the space of streets, squares, panoramas, city gardens, etc.),
2) compositional (regarding the exposure of panels in architectural form of the buildings, in the compositions of objects façades).
The methodology for assessing the impact of selected photovoltaic investments on Bialystok and Cordoba cityscape was devoted on the basis of selected elements of the view analysis methodology by Forczek-Brataniec [16] and "7S" method by Ozimek [17], representing the so-called The Cracow School of landscape Architecture (Cracow University of Technology, Poland) [18] and the own findings of interdisciplinary team of authors [19, 20]. The conclusions of the research on impact of tall buildings [21, 22] and large investments [22, 23, 24] on the cityscape were also taken into account.
Forczek-Brataniec [16] dedicates her original view analysis to the assessment of the impact of the planned investments on the landscape. She distinguishes between three basic object models: point, line and area. She proposes a methodology for assessing these objects in the landscape through the analysis of active and passive visibility in terms of the potential, real and projected view range. The method was developed for the purpose of assessing the impact of the investment on a large area, e.g. wind farms (a group of wind towers), or road investments on the non-urban landscape, including a protected landscape with significant natural or cultural values. Based on this methodology, the categories of point and surface models as well as their active exposure were adapted to the needs of the study, due to their adequacy to the scale of the examined objects and their visibility in the landscape. The actual view range of the examined objects was assessed, i.e. the real view range determined on the basis of the topography and land cover, due to the location of the examined objects in the urban landscape. The farthest distance from which the examined objects are visible was assumed as the main linear parameter of the view range assessment. It is the result of the urban structure of cities.
The "S7" Computer Parameterization of View Components was developed by Ozimek [17]. It enables the use of digital techniques (numerical indicators) in order to objectify the assessment of the impact of the investment on the landscape. It is applicable to both existing and planned facilities. The method recommends expert selection of the features of the object under study. These are: colour, shape, surface area, height, percentage of the view, interference with the sky line, view structure. Among the above features, the study used criteria with specific categories: colour (determined by contrast or inscribed in the background), shape (harmonious, neutral, disharmonious), surface area (only for surface objects on the ground), height (of the object on which the panels were placed solar panels or collectors) and interference with the sky line (only for panels and collectors).
The study was supplemented with criteria related to the scale: building - photovoltaic panels or collectors. The location of the panels in relation to the building (on the ground, wall, roof), integration of the panels into the architectural form of the facility (harmonious, indifferent, disharmonious) and the location of the facility in the city landscape (architectural and landscape interior: street, square, open interior, green surroundings, panorama) were classified.

3. Results

3.1. Cordoba (Spain)

In Cordoba, 4 locations of solar panels and solar collectors characteristic for this city were analysed. The summary (Table 1) of the analysed data allows to determine that the panels were located both in the context of panoramic views and street interiors. It should be emphasized that these solutions have a positive or neutral effect on the exposure of the panorama. The surface realization of roofs over car parks (Table 1: C3), although designed and installed 20 years ago, can still be assessed as positive in terms of landscape. Its surface-linear character harmonizes with the wide, horizontal, open landscape.

Table 1. Analise of photovoltaic panels and solar collectors location in cityscape of Cordoba

	Photo of location	Type of cityscape view	Point/ area object	Location on the building	Real Zone of Visual Influence (m)	Shade (in relation to the background)	Shape (in relation to the building form)	Surface area (m2)	Heinght/ size (m)	Silhouette/Sky line	Composition	Assessment
Cordoba, Menindez Pidal street												
C1		panorama	area	on the groud	150-250	contrast	harmonious	15000	3	0	1	P/NEU
Cordoba, city center												
C2 A		street	point	roof	20	contrast	neutral	-	12	1	0/1	NEU
Campus Rabanales, st. Colonia San José street												
C2 B		street	point	roof	20	contrast	neutral	-	12	0	0/1	NEU
Campus Rabanales, St. Astrónomo Kepler street												
C3		street, panorama	area	on the ground	50-75	blending	harmonious	750	3	1	1	P/NEU

The photovoltaic farm (Table 1: C1) located on the slope of the valley on the A-3050 road to Cordoba is inscribed in the background of the view, below the buildings of the medical university. The strip layout of the power plant converts to the horizontal composition of the view.
Photovoltaic panels and solar collectors are placed on the roofs of apartment buildings. Their visibility is largely limited due to the small area of actual visibility. The analysed investments (Table 1: C2A, C2B) do not degrade the city's view.

3.2. Bialystok (Poland)

In Białystok, 10 locations of photovoltaic panels and solar collectors on buildings and in their surroundings were analysed. The list (Table 2) shows that the most characteristic type of architectural and landscape interior, in which the panels are located, is the street. On the street scale, they are visible in buildings with various functions: residential, service, and public. The applied solutions have a positive, neutral and negative impact on the city landscape.
The degrading influence of a mini solar farm (Table2: B1) on the street landscape was established due to the disharmony with the neighbouring buildings. Technological solutions in this form should not be located in an urbanized city centre, in the context of an intense urban and architectural fabric.
The building facades are a particularly aesthetically sensitive surface for placing photovoltaic panels. The indicated examples (Table 2: B2A, B2B) confirm that the implementation of the investment should be carried out with care to preserve the compositional values of the building's body. The lack of harmony between the newly placed elements of the photovoltaic installation and the facade (Table 2: B2B) leads to the degradation of the aesthetics of the facility, and thus has a negative impact on the city landscape. In solutions assessed as positive or neutral for the city landscape, new technological elements do not destroy the existing architectural composition of the building.

IOP Conf. Series: Earth and Environmental Science 1006 (2022) 012011 doi:10.1088/1755-1315/1006/1/012011

Table 2. Analise of photovoltaic panels and solar collectors location in cityscape of Bialystok

	Photo of location	Type of cityscape view	Point/ area object	Location on the building	Real Zone of Visual Influence (m)	Shade (in relation to the background)	Shape (in relation to the building form)	Surface area (m2)	Heinght/ size (m)	Silhouette/Skyline	Composition	Assessment
	Bialystok, Kalinowskiego street											
B1		street	point	on the ground	80-50	contrast	**degrading**	150	3	0	-1	**NEG**
	Bialystok, Skladowa street											
B2A		street	point		120-60	contrast	neutral		6	0	+1	NEU
	Bialystok, Wiejska street											
B2B		street	point		300	contrast	**degrading**		9	0	-1	**NEG**
	Bialystok, Legionowa street											
B3A		street	point		200	blending	harmonious		12	0	1	P/NEU
	Bialystok, Wiosenne street											
B3B		street	point		150	blending	harmonious		12	0	+1	P
	Bialystok, Kawaleryjska street											
B3C		street	point		40	blending	harmonious		6	0	+1	P
	Bialystok, Zielna street											
B3D		street	point		25	contrast	**degrading**		6	-1	-1	**NEG**
	Bialystok, Zwierzyniecka street											
B3E		street, panorama	point		550	contrast	neutral		30	+1	1	P/NEU
	Bialystok, Kosciuszko Square											
B4A		square	point	on the ground	20	contrast	neutral		1	0	0	P/NEU

				Bialystok, Central Garden								
B4B		garden	point	on the ground	20	contrast	neutral		1	0	0	NEU

The location of the panels on the roofs can lead to positive, neutral and negative solutions for the landscape. In the projects assessed as positive, photovoltaic installations were placed on the roofs, harmonizing with their shape and colour (Table: B3A, B3B, B3C). In the examples that are destructive to the landscape, we usually deal with solutions where panels or collectors, for climatic and shading reasons, are placed in a way that contrasts with the shape of a single-family house (Table 2: B3D). The small size of the building affects its visibility on the street scale. As a result, technical elements that are dissonant with the architecture of the building, interfering with the sky line, also have a destructive effect on the view of the street / city landscape. Panels placed on flat roofs of high-rise buildings (Table 2: B3E) have a neutral effect on the landscape due to the considerable distance of their observation.
Architectural furniture such as benches with photovoltaic panels deserve consideration when locating them in the city space (Table 2: B4A, B4B). The examples from Bialystok indicate that mistakes are made in the orientation of the benches with respect to the directions of their insolation (Table 2: B4B). The panels are exposed to acts of vandalism due to their too low location (Table 2: B4B).
Photovoltaic panels and solar collectors are placed on the roofs of apartment buildings. Their visibility is largely limited due to the small area of actual visibility. The analysed investments (Table 1: C2A, C2B) do not degrade the city's view.

4. Discussion

In Bialystok and Cordoba, photovoltaic panels and solar collectors have an impact on the city's landscape. In Cordoba, more favourable solutions have been applied in terms of placing photovoltaic panels on the ground. It should be emphasized that they are integrated into the city with their location on wastelands and outside the city centre.
Placing a photovoltaic system on roofs affects the landscape in different ways: positively, indifferently and negatively. Relatively numerous examples of landscape degradation in this respect can be found in Białystok, especially in relation to single-family housing. The applied solutions do not harmonize with the architecture in term of form. Schuetze already drew attention to the integration of the form of photovoltaic panels with the body of the building in terms of aesthetics and technology [14]. The author called for the creation of a support policy in this direction. So far, administrative activities in Poland (Bialystok) have been limited to financial programs supporting investments, without taking into account its impact on the city landscape. The financial factor understood as the minimum investment cost often affects the color disharmony between the panels used and the roofs and facades. Many investments in the Białystok region are subject to destruction, despite the fact that the available technological solutions and numerous examples prove the possibility of composing and harmonizing the colour of facades and roofs [4, 6].
So far, examples of the use of photovoltaic panels on walls are typical of Bialystok. They concern service facilities and administration. Among them, we can also find positive and negative projects for the city landscape. Investments in Białystok are characterized by the use of one form and colour of photovoltaic panels. Examples, ie Powerhouse Brattørkaia (NO), Umwelt Arena (CH) or Copenhagen International School (DK) discussed by Devetakovi´c et al. [15] confirm much wider formal and colour possibilities of photovoltaic systems that can be used on facades. BIPV Systems such as tiles, modules, glazing or foil offer wide possibilities of harmonious design of Building-Integrated Photovoltaics [10].
Attention is drawn to the fact that photovoltaic panels in the examined cities Bialystok and Cordoba are most often visible on the scale of streets and selected sections of panoramas. Their influence on the landscape of squares was not observed. The diagrams proposed by Celadyn and Filipek [6] and Gawryluk and Krawczyk [20] in relation to the horizontal projection and Levels of close and remote visibility of PV surfaces from public domain reproduced by Devetakovi´c et al. [15] with permission from Maria Cristina Munari Probst, Christian Roecker [25]. The use of the above-mentioned patterns

would serve to protect the values of the city's landscape in the scale of streets and panoramas. It should be emphasized that photovoltaic installations are not mounted on monuments in Bialystok and Cordoba yet. The issue of using photovoltaic panels on monuments is discussed by Tsoumanis et al in the context of the Smart Evolution of Historical Cities understood as Integration of Innovative Solutions Supporting the Energy Transition while Respecting Cultural Heritage in a case study of Evora (Portugal) [11]. Many projects to date are discussed due to implementation of somewhat controversial photovoltaic systems on heritage buildings. Rosa proposed indications and cues in evaluating the limits that RES solutions, and in this case BIPV, can present on the historical building stock [12].
Due to the location of solar farms in the context of Bialystok and Cordoba cities, the conducted research should be continued with regard to the bird's eye view of the city.

5. Conclussion

Photovoltaic panels and solar collectors have an impact on the landscape of the examined cities in the street and panorama scale. The conclusions from the analysis of the solutions used in Bialystok and Cordoba show that solar farms should be used in urban wastelands outside the city centre. Moreover it was presented that the use of panels on roofs and walls affects the city landscape. Negative solutions result from aesthetic disharmony in terms of colour and shape in relation to the form of the building. Solutions harmonized with the shape and colour of the building can be positive or neutral to the landscape and at the same time optimal in terms of efficiency. Finally, it is worthy to note that in order to avoid the degradation of the city landscape, an interdisciplinary team should be involved in the process of designing the location of solar panels and solar collectors: an installation designer, architect or landscape architect.
Analysis of the impact of photovoltaic panels and solar collectors (existing or planned) based on the criteria presented in our paper can be used to assess and take appropriate design decisions that are right for the city landscape. The proposed applications can be used in the planning, design and investment process in Bialystok and Cordoba, as well as in other cities.

6. References

[1] Velasco A and García S J and Guardo A and Fontanals A and Egusquiza M 2017 *Assessment of the Use of Venetian Blinds as Solar Thermal Collectors in Double Skin Facades in Mediterranean* Climates. Energies 10(11) p 1825; https://doi.org/10.3390/en10111825

[2] Bandaru S H and Sourav V B and Radulovic K J and Hutchinson D and Khusainov R 2021 *A Review of Photovoltaic Thermal (PVT) Technology for Residential Applications: Performance Indicators, Progress, and Opportunities* Energies 14(13) p 3853 https://doi.org/10.3390/en14133853 - 26 Jun 2021

[3] Abdelhafez MHH, Touahmia M, Noaime E, Albaqawy GA, Elkhayat K, Achour B, Boukendakdji M 2021 Integrating Solar Photovoltaics in Residential Buildings: Towards Zero Energy Buildings in Hail City, KSA. *Sustainability*. 13(4):1845. https://doi.org/10.3390/su13041845

[4] Visa I and Comsit M and Duta A 2014 *Urban acceptance of Façade Integrated Novel Solar Thermal collectors* Energy Procedia 48 pp 1429-1435

[5] Pelle M, Lucchi E, Maturi L, Astigarraga A, Causone F 2020 *Coloured BIPV Technologies: Methodological and Experimental Assessment for Architecturally Sensitive Areas*. Energies 13(17):4506. https://doi.org/10.3390/en13174506

[6] Celadyn W and Filipek P 2020 *Investigation of the Effective Use of Photovoltaic Modules in Architecture*. Buildings 10(9):145. https://doi.org/10.3390/buildings10090145

[7] Horváth M and Kassai-Szoó D and ,Csoknyai T 2016 *Solar energy potential of roofs on urban level based on buildings typology* Energy and Buildings 111 pp 278-289

[8] Takebayashi H and Ishii E and Moriyama M and Sakaki A et al. 2015 *Study to examine the potential for solar energy utilization based on the relationship between urban morphology and solar radiation gain on building rooftops and wall surfaces* Solar Energy 119 pp 362-369

[9] Aimilios M and Bougiatioti F and Oikonomou A 2010 *Less could be more: Architectural integration of active solar systems in existing urban centres* 7th Mediterranean Conference and

Exhibition on Power Generation, Transmission, Distribution and Energy Conversion 7-10 November 2010, Agia Napa, Cyprus (Paper No. MED10/189)

[10] Gholami H and Nils Røstvik H and Steemers K. 2021 *The Contribution of Building-Integrated Photovoltaics (BIPV) to the Concept of Nearly Zero-Energy Cities in Europe: Potential and Challenges Ahead* Energies 14(19):6015. https://doi.org/10.3390/en14196015

[11] Tsoumanis G and Formiga J and Bilo N, Tsarchopoulos P and Ioannidis D and Tzovaras D 2021 *The Smart Evolution of Historical Cities: Integrated Innovative Solutions Supporting the Energy Transition while Respecting Cultural Heritage.* Sustainability 13(16):9358. https://doi.org/10.3390/su13169358

[12] Rosa F 2020 *Building-Integrated Photovoltaics (BIPV) in Historical Buildings: Opportunities and Constraints.* Energies 13(14):3628. https://doi.org/10.3390/en13143628

[13] Radovanovic I and Popovic I 2021 *Identification of Degrading Effects in the Operation of Neighboring Photovoltaic Systems in Urban Environments.* Electronics 10(7):762. https://doi.org/10.3390/electronics10070762

[14] Schuetze T 2013 *Integration of Photovoltaics in Buildings—Support Policies Addressing Technical and Formal Aspects.* Energies 6(6):2982-3001. https://doi.org/10.3390/en6062982

[15] Devetaković M and Djordjević D and Radojević M and Krstić-Furundžić A and Burduhos B-G, Martinopoulos G and Neagoe M and Lobaccaro G 2020 *Photovoltaics on Landmark Buildings with Distinctive Geometries.* Applied Sciences 10(19):6696. https://doi.org/10.3390/app10196696

[16] Forczek-Brataniec U 2018 *Visible space. A visual analysis in landscape planning and design* (Kraków: Wydawnictwo PK)

[17] Ozimek A 2019 *Measure of the landscape. Objectification of view and panorama assessment supported with digital tools* (Kraków: Wydawnictwo PK)

[18] Ozimek P and Bohm A and Ozimek A and Wańkowicz W 2013 *Planning spaces with high scenic values by means of digital terrain analyses and economic evaluation* (Kraków: Wydawnictwo PK)

[19] Krawczyk D and Kolendo Ł 2016, *Wykorzystanie GIS i danych klimatycznych do obliczeń potencjału solarnego możliwego do wykorzystania w systemach C.W.U. i C.O.* XVI Konferencja Naukowo–Techniczna : Rynek Gazu 2016 : XXII Konferencja Naukowo–Techniczna : Rynek Ciepła - REC 2016 : Materiały Konferencyjne [Dokument elektroniczny]. Redakcja Rynku Energii. pp 63–69

[20] Gawryluk D and Krawczyk D A 2021 Photovoltaics and architecture in contemporary cityscape *Future of the City* ed D Gawryluk and D A Krawczyk (Bialystok: Oficyna Wydawnicza PB) chapter 6 pp 73-82, DOI: 10.24427/978-83-66391-62-8

[21] Czyńska K 2017 *Assessment of Tall Buidings Visual Impact on Selected Landscape Interiors Using the VIS Method* Space & Form 32 pp. 237-256

[22] Czyńska K and Rubinowicz P and Zwoliński A 2017 *Analyses of tall buildings in the cityscape,* Teka Komisji Urbanistyki I Architektury PAN/ Oddział w Krakowie, vol. XLV, pp. 319-341

[23] Orzechowska-Szajda I 2018 *Visual Landscape Inventory (VLI) as a tool supporting landscape planning and management* Space & Form 33 pp. 225-252

[24] Sas-Bojarska A 2017 *Big-scale activities in the context of landscape threats and protection* (Gdańsk: Wydawnictwo Politechniki Gdańskiej)

[25] Munari Probst M C and Roecker C 2019 *Criteria and policies to master the visual impact of solar systems in urban environments: The LESO-QSV method.* Solar Energy (Elsevier) 184, 672–68

Acknowledgments
This study and publication was financed within the project BUT InterAcademic Partnerships (PPI/APM/2018/1/00033/U/001). Additionally it was conducted within scientific cooperation “The possibility of the renewable energy sources usage in the context of improving energy efficiency and air quality in the buildings and civil constructions” between BUT and UCO and the science research funds at Bialystok University of Technology WZ/WB-IIS/9/2019, as well as WZ/WA-IA/6/2020.

IOP Conf. Series: Earth and Environmental Science 1006 (2022) 012012 doi:10.1088/1755-1315/1006/1/012012

The Impact of Urban Green-infrastructure Development on the Price of Surrounding Real Estate: A Case Study of Taichung City's Central District

Kuo-Wei Hsu[1] and Jen-Chih Chao[1]

1 Department of Landscape and Urban Design, Chaoyang University of Technology, Taiwan

Email: kwhsu@cyut.edu.tw, jenchihchao@gmail.com

Abstract. Amid increasingly common extreme climate events and urban heat-island effects, cities are at increased risk from a range of natural disasters. Green infrastructure emphasizes coordination and mutual benefit between, on the one hand, human construction and development, and the natural environment on the other, and can be adjusted to mitigate disaster risks. However, in addition to public-sector policies, urban green-infrastructure development requires incentives for private-sector implementation. This study examines the impact on the value of local real estate of three green-infrastructure projects in the Central District of Taichung City, Taiwan, both during and after their construction. Its findings show that green-infrastructure development can indeed drive-up urban real-estate prices, with slightly different rates of increase, according to type of real estate themselves. The knowledge of this dynamic provided by this study should make such development attractive to the private sector, and thus ultimately aid disaster prevention and improve the overall quality of the urban environment.

1. Introduction

As a result of climate change caused by global warming, Taiwan's urban areas are facing various threats, including heat-island effects, flooding, and fires. These changes are having impacts on the natural and built environments alike. Conceptually, green-infrastructure planning emphasizes linking landscape and ecology with urban/community life and space utilization, and implementing the results into sustainable urban land-use planning (Benedict & McMahon, 2006; Brown, 2008) [1-2].It is expected that urban development and warming will both intensify in the coming years (e.g., Norton et al., 2015, Watkins et al. 2007, Papanastasiou et al. 2010, Argüeso et al., 2014)[3-6], but scholars including Matthews et al. (2015) believe that green infrastructure can play a potentially important role in adapting to some of the expected impacts [7]. Adoption of green infrastructure can preserve more natural-ecosystem functions and values, while also providing various benefits to humans (Benedict & McMahon, 2002) [8]. Groot (2006) proposed that to reconcile landscape protection with changing land-use and natural-resources needs, the ecological, socio-cultural and economic values of the landscape must be fully considered in planning and other forms of decision-making [9]. ECOTEC (2008) reported that the economic benefits of green-infrastructure investment are both direct and indirect, and both quantifiable and non-quantifiable, given the role of public- and private-sector cost reduction and risk-management effects [10]. This reflects that green infrastructure has a wide range of impacts, and that its environmental, social, economic and ecological ones are all magnified by climate change.

Green infrastructure can be defined as a network linking natural areas and other open spaces, providing air and water purification, promoting the movement of species and preserving biological habitat, and generally serving as an important framework to support the operation of the natural environment in the

IOP Conf. Series: Earth and Environmental Science 1006 (2022) 012012 doi:10.1088/1755-1315/1006/1/012012

urban (Benedict & McMahon, 2006) [1]. In recent years, Taichung City, Taiwan, has invested heavily in the development of green infrastructure in its the Central District to improve the urban environment and build disaster preparedness and resilience. The purpose of this study is to examine the impact this green-infrastructure drive has had on real-estate prices. It is expected that the effectiveness of public-sector construction will boost demand in the private sector, and that the benefits of green infrastructure will be accumulated in a point-line manner, so as to accelerate the construction of urban green-infrastructure network and thus reduce negative climatic impacts.

2. Methods

This study uses a case-study approach to examine the implementation of three green-infrastructure projects – the Taichung Overpass Project, the Hsin-Sheng Green River Corridor Project, and the Liuchuan Blue Belt Riverfront – all of which are located in Taichung City's Central District. The impact of green-infrastructure development on real estate prices will be assessed from the planning and design stages of these projects through to the completion of construction.

2.1. The Liuchuan Blue Belt Riverfront

The main reason for the poor water quality in Liuchuan is upstream sewage, and in particular, the direct discharge of sewage from both sides of the river. The first step of this remediation project is to intercept sewage to reduce pollution at the source, and to set up an underground gravel contact-aeration water-purification system. The intercepted effluent will be treated and released on site to be used as the water source for a low-impact development demonstration section, and at the same time, the flood-control capacity of the channel will be strengthened and a flood-warning mechanism established to improve the safety of public recreation. The wider aim is the same as for the Hsin-Sheng Green River Corridor Project. Construction began in March 2014 and was completed in October 2016.

2.2. The Hsin-Sheng Green River Corridor Project

The Green River project is divided into three components: water purification, water safety, and water space. The first involves intercepting polluted water from upstream, bringing in fresh water from streams to supplement that in the Green River, and using inter-gravel treatment to achieve water purification. The second focuses on flood prevention and control, with flood-control facilities configured to ensure the strength of shorelines, enhance protection standards, and generally provide a safe water environment for the public. Lastly, the water space component aims to create a unique aquatic and green open space in the Green River by restoring habitats, recreating the river's natural ecology, and integrating the rich human history of the surrounding area, in such a way that the day and night have different looks. The wider purpose of this is to transform the surrounding residential and commercial space, and bring back the prosperity of the Central District. Construction began in October 2016 and was completed in March 2018.

2.3. The Taichung Overpass Project

With the new train station as its core, the Central District will become a "cultural city within a city" in Taichung, with a diversified city art museum and green museum. As well as being green and cultural, it will be water-friendly and near water, with "green + railway heritage + green river" describing its overall spatial-development vision. The old 1.6 km railway line from the Taiwan Sugar factory has been used to improve the space for pedestrians and bicycles, and as an axis for the development of the Central District, linking the Railway Cultural Park, the Martial Arts Hall, the Brewery, and other scenic spots with the Zhongzheng Road, Jianguo Market, and the back station shopping area. The "lines" are used to drive the development of various "points", providing people with the opportunity to roam around the scenic spots along the former railroad tracks on foot and by bicycle, as well as complementing the improvement of the Green River and strengthening public facilities in the inner city. Reuse and development of important historical buildings (including private properties) and their surrounding land has created opportunities for the co-innovation of the inner city's historical district image with the non-riverside areas and the land of Taiwan Sugar, and to drive the development of the Central District

IOP Conf. Series: Earth and Environmental Science 1006 (2022) 012012 doi:10.1088/1755-1315/1006/1/012012

through enhanced public transportation, low-carbon transportation, history, and industrial culture and creativity. Construction began in 2018 and was completed in 2020, as show in Figure 1.

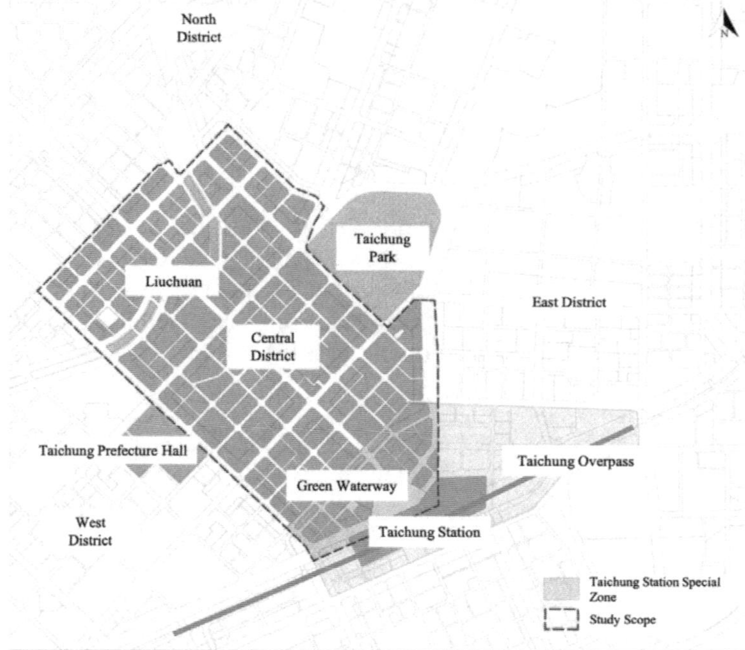

Figure 1. Distribution of green-infrastructure projects in Taichung's Central District.

3. Results and discussion

The relative positions of the Taichung Overpass, Green River, and Liuchuan River are shown in Figure 1. This study's examination of changes to land and real-estate prices in the study area was conducted in the first instance through inquiries to the Ministry of the Interior, which established that green-infrastructure construction indeed coincided with a local upsurge in land and other real-estate transactions. Analysis of data gathered from the Houseplus website and other real-estate websites suggested that this increase was non-significant.

From the above trend charts of real-estate transaction volumes and prices in Central District, it can be seen that the initial promotion of green-infrastructure planning and design in 2014 coincided with the beginning of a slow rise in prices. Due to the nature of the current real estate market in the Central District, the owners of many old commercial buildings have been expressing their desire to start anew, as there are no new construction projects. In addition to the old Far Eastern Building, which has been declared a landmark building in the district, an Urban Renewal Association has also been organized, branches of which have been established for the Chien Yue Building, the Golden Garden Building and the Man Shing Building, showing how green infrastructure can trigger urban renewal. In addition, the number of buildings renovated by the private sector and their values have increased markedly. In view of the slow pace of urban renewal, the private sector has improved the quality of the urban environment and enhanced the value of real estate through additions and renovations, as shown in Figure 2.

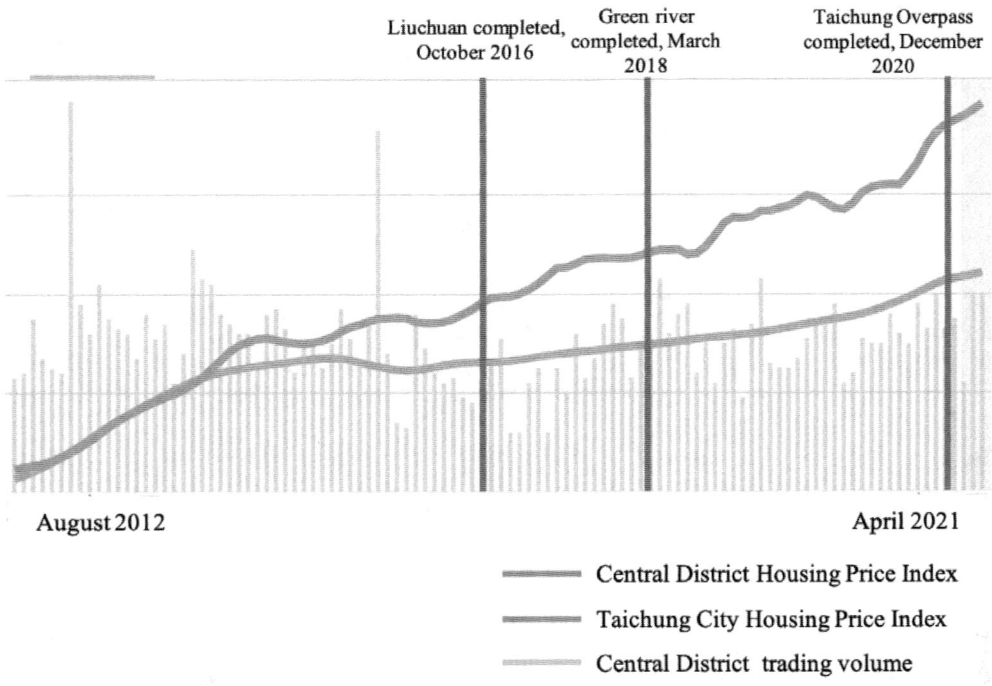

Figure 2. Trends in real estate prices and trading volume in central Taichung.
Source: https://www.houseplus.tw.[11]

Noelwah et al. (2014) found that the sale prices of real estate increased with their distance from low-impact development facilities [12]. The city of Portland, Oregon, has a long history of promoting low-impact development facilities and continues striving to improve its air and water quality, increase pedestrian and bicycle access and safety, expand green spaces and wildlife habitats, and boost real estate values through its Green Streets program (City of Portland Environmental Services, 2008) [13]. Nazir et al. (2015) showed that green-infrastructure elements bring economic, social and environmental benefits to the community, and positively demonstrates the indirect contribution of green infrastructure to house and other real-estate prices [14]. The results of this study confirm these previous studies' findings in the case of Taichung, suggesting that more such infrastructure will further benefit both the public and private sectors.

4. Conclusion

Urban green infrastructure is aimed at improving the quality of the urban environment, in part by enhancing its green space, and mitigating the effects of extreme weather and climate, as well as urban heat islands. Prior studies of green infrastructure development suggest that the value of such projects to their local areas' ecology, human health and well-being, and recreation are factors that influence purchases of real estate and the prices paid for it. Based on this study's examination of the impact of green-infrastructure development on real-estate values, its authors expect more public- and private-sector actors to invest in such infrastructure's construction. It is to be hoped that this will eventually lead to the establishment of urban green-infrastructure networks, which will foster friendly and comfortable urban environments in Taichung and beyond.

5. References

[1] Benedict M A and McMahon E T 2006 Green Infrastructure: Linking Landscapes and Communities Island Press Washington
[2] Brown P R 2008 The Role of Citizen Activists in Urban Infrastructure Development. In Eugenie L. and Susan M. Wachter （eds.）, Growing Greener Cities: Urban Sustainability in the Twenty-First Century. Philadelphia PA: University of Pennsylvania Press 152–169

[3] Norton A B, Andrew M, Coutts M A, Livesley J S, Harris J R, Hunter M A and Williams S G N 2015 Planning for cooler cities: A framework to prioritise green infrastructure to mitigate high temperatures in urban landscapes Landscape and Urban Planning **134** 127–138

[4] Watkins R, Palmer J and Kolokotroni M 2007 Increased temperature and intensification of the urban heat island: implications for human comfort and urban design Built Environment **33(1)** 85–96

[5] Papanastasiou D K, Melas D, Bartzanas T and Kittas C 2010 Temperature, comfort and pollution levels during heat waves and the role of sea breeze International Journal of Biometeorology **54(3)** 307–317

[6] Argüeso D, Evans J P, Fita L and Bormann K J 2014 Temperature response to future urbanization and climate change Climate Dynamics **42(7–8)** 2183–2199

[7] Matthews T, Lo Y A and Byrne A J 2015 Reconceptualizing green infrastructure for climate change adaptation: Barriers to adoption and drivers for uptake by spatial planners *Landscape and Urban Planning* **138** 155–163

[8] Benedict M A and McMahon E T 2002 Green Infrastructure: Smart Conservation for the 21st Century. Renewable Resources J. **20(3)** 12-17

[9] De Groot R 2006 Function-analysis and valuation as a tool to assess land use conflicts in planning for sustainable, multi-functional landscapes Landscape and Urban Planning **75** 175-186

[10] ECOTEC Research and Consulting for NENW 2008 The Economic Benefits of Green Infrastructure: developing Key Tests for evaluating the benefits of Green Infrastructure

[11] House+ website Available Online: https://www.houseplus.tw. (Accessed on June 2021)

[12] Netusil N R, Levin Z, Shandas V, and Hart T 2014 Valuing green infrastructure in Portland, Oregon. Landscape and Urban Planning **124** 14-21

[13] Environmental Services City of Portland 2008 Stormwater management for clean rivers: Green Streets

[14] Nazir N N M, Othman N and Nawawi A H 2015 Role of green infrastructure in determining house value in Labuan using hedonic pricing model Procedia-Social and Behavioral Sci **170** 484-493

Sustainable Cities, Rescue of Original Construction Methods and Use of Technology. A literary Review

Susana Moya Vicuña[1*]

1Faculty of Architecture, Arts and Design (FAAD), Indoamérica Technological University, Sabanilla Street, Quito 170103, Ecuador
E-mail: susanamoya@uti.edu.ec

Abstract. The concept of sustainability and even more so applied to the city is quite broad, however in general we could say that they are those whose planning and growth considers current challenges, but also understands and recognizes those next to appear, such as the upsurge of climate change, the growth and migration of the human population in and towards cities, based on which it designs a model based on economic and social development, thinking about the importance of people and their needs. In recent years, several sustainable cities have appeared that are proposed as a model, many of these cities have not only looked to the future with new technologies that offer us efficient systems, but have also looked to the past and have found vernacular architecture of their environments, practical solutions of a passive nature, however, although these model cities can serve as an example and even as a hotbed for the development of techniques and technologies that lead us to a sustainable city, they are not models that can be adapted in an intact way in any territory, but must be reinterpreted to achieve optimal functioning among different conditions of another territory.

1. Introduction

During the years after industrialization, the conception of buildings and therefore their design and construction, was based on the use of fossil fuels to maintain comfort, this has resulted in the existence of buildings and urban areas whose characteristics do not have relationship with the environment, and they have been "exported", proliferating an architecture that does not take into account the surroundings environmental conditions, and does not take advantage of climatic conditions to generate comfort and energy efficiency, requiring a large amount of energy to function properly.

However, in current times new rules have been stablished, the demographic dynamics of the population, which has practically tripled since the 50s of the last century until now, and has not stopped but continues increasing, and whose mobility pushes on cities, since around 60% of the world population will live in cities by 2030, according to ONU habitat in 2018, for most of today's cities this population growth, together with a capitalist model based on consumption, in which a huge number of people depend on fossil fuels, and which works under a linear approach to urban metabolism, in which the city is only a producer of waste while consuming the resources of other areas, cities currently consume 80% of the world's energy, and more than 60% of the resources, while producing around 70% of the world's carbon emissions; despite the fact that they cover less than 3% of the earth's surface, the construction sector is responsible for 40% of CO_2 emissions, 30% of raw material consumption, 20% of water consumption, 30% of waste generation and a significant part of land occupation [1, 2, 3, 4, 5].

These characteristics not only cause an impact on the environment and the health of citizens, but also cause social inequality that is reflected in access to properly planned environments that ensure infrastructure and services for the population, since normally people with low purchasing power who settle on the outskirts of cities, increasing inequality and the risk of this population living in unhealthy areas [2, 3]

The current situation in several cities, soon the level of pollution and resource hoarding will be unsustainable, unless another model is started. This has caused the appearance of cities whose planning is based on sustainability and are presented as a new city model, which throughout its design and construction stage contemplate and implement various innovations, through the use of technologies of latest generation, as well as the reuse of construction materials, the use of renewable energies, means of transport and ecological materials, among others such as their construction methods and space design; in order to build a self-sufficient city working under a circular metabolism model [2, 3]

2. Methodology

The development of this research is based on the collection and bibliographic analysis of information available in scientific articles, books, magazines, newspapers and web pages. The cities of Masdar and Tianjin are analyzed, which together with Dongjiang between 2007 and 2008 were fighting to become the first sustainable city built in the world from scratch.

These two cities intend to impose themselves as a model to follow, with the corresponding climatic adaptations of each environment, their similarities are several, but they present differences in many occasions marked not so much by the climatic characteristics of their surroundings, as by the characteristics of their population needs, which makes the model that each poses present different perspectives consolidated in the use of intelligent technologies in the case of Tianjin, and a mixture of technology and tradition in Masdar that recovers and rethink strategies that ancient cities were built and that had been left behind by the rise of technology and the ease of achieving comfort through the indiscriminate use of energy, reinterpreting old models that merge with the innovations of the present to rise to a new city model, based on sustainability.

3. Sustainable cities

Sustainability concept and even more applied to the city is quite broad, it refers not only to the set of planning activities for growth and infrastructure of the city, but also encompasses political and sociocultural environment, having its own characteristics depending on the needs of each urban environment, and the issues that each one wishes to focus on its development, since as well as urban planning itself, the configuration of a sustainable city is a transdisciplinary work [5, 6].

In general, sustainable cities are those whose planning and growth considering the challenges we currently face, but also understand and recognize those next to appear, such as the worsening of climate change, the growth and migration of the human population in and towards cities, designing a model based on economic and social development, thinking about the importance of people and their needs [5, 7]. But it also raises a life and urban growth in harmony with the environment, aware of the serious consequences that they can bring, the use of fossil fuels, polluting materials, reduction of green spaces and endemic biodiversity, affecting the quality of life of all the inhabitants on earth with consequences such as climatic factors disorders, considering that vegetation regulates CO_2, and all these factors function as a chain in which a small variation in one of it, produces alterations in all of them. [8, 9, 10]

The sustainable city is a city that analyzes and takes advantage of the characteristics of its environment through the formation of an urban plan that introduces "innovative" elements such as the use of construction methods, materials and characteristics of the original or native buildings of the environment where it is implanted as well as new technologies that allow reaching goals previously unthinkable and that in most cases allow a circular metabolism of the city, promoting a change in urbanization form, in which not only modern technology is used but also the native strategies of the environments are rescued and enhanced, designed to function and generate comfort without damaging the environment or consuming large amounts of fossil fuels [6, 7, 11].

3.1. Sustainable cities in the world

In the future several cities around the world will suffer serious impacts due to climate change, decreasing the quality of life of their inhabitants and they will surely feel with greater force in developing countries, therefore, it is necessary that new cities are planned and developed under the parameters of a sustainable city, within one of the most important aspects in the design section is the adaptation to the environment, based on its climate, materiality resources, strategies and socio-cultural

characteristics, which will influence the shape, distribution and optimal functioning of the city. In recent years, several sustainable cities have appeared that are proposed as a model. Many of these cities have not only looked to the future with new technologies offering efficient systems but have also looked to the past and have found vernacular architecture of their environments, practical and passive solutions that, on many occasions, have been enhanced with technologies, but which are still solutions designed and used based on the understanding of their own environment. [12, 13]
However, although these model cities can serve as an example and even as a hotbed for the development of techniques and technologies that lead us to a sustainable city, they are not models that can be adapted intact in any territory on the planet, but rather they must be reinterpreted to achieve optimal functioning in different conditions of another territory, ranging from its climate, idiosyncrasies and people customs [12, 13]

3.2. Masdar. Technology and tradition
Masdar appears in a period in which sustainability has become a necessity to maintain the resources that we still have, just as the first "city" made its appearance also in the East, at the end of the Neolithic, in response to their communication needs, product exchange and community life.
Located in Abu Dhabi and designed by Foster and Partner research, *"the city combines cutting-edge technologies with the principles of planning traditional Arab settlements"* [14], so that it responds, specifically, at its location and climate.
The design approach of the city focused on certain points that governed urban design, such as a compact urban form that allows to connect the spaces in such a way that the routes that the inhabitants must make to reach the different points of the city are short, thus avoiding the intensive use of vehicles that are in fact electric, in addition having an excellent public transport service, also electric, with a planned population that is relatively homogeneous in terms of socioeconomic characteristics (50,000 permanent residents and 40,000 temporary residents in 6.000 km), the neighborhoods were raised from the conceptual stage to adopt the traditional approach, where the framework of the city is inspired by the traditional Arab city, of mixed use, with narrow and shady streets, low height despite the high density, using traditional materials that manage the temperatures of the environment, providing a more balanced and fair community, through well-distributed urban spaces and squares [13, 14]
One of the most remarkable architectural "innovations" of Masdar is in its design, a mixture of tradition and modernity, in which elements of the vernacular construction of the East and current technology are combined, reinterpreting knowledge of times when it was not thought of energy expenditure but which, nevertheless, was minimal since its architectural elements, as well as its materials, were a response to the location, the climate and the needs of each region [13, 15]
In ancient times, in the cities of the East the Mosque was the initial design point, from which the city was organized, in Masdar "The Science and Technology Masdar Institute" is one of its most relevant points, around which the rest of the spaces of the city are designed with an orientation from northwest to southwest, connected by a network of narrow streets that connect with parks and squares, the design streets inspired by the ancient architecture of the East, in which adobe houses and buildings do not exceed the 5 floors, give a greater shade along the journey through the narrow streets, to get a microclimate that favors the air movement, intending to reduce at 20 ° C the internal temperature of the city, generating a comfortable climate between 20 or 25 ° C in an area whose temperatures can reach around 45 ° C, and implementing this design based on the old Masdar measures, it provides a response to these sunny conditions, also has a cooling system by water sprinklers, to maximize the results of the vernacular design implemented, connecting in this way modernity and tradition by using technology in such an old passive design strategy [13, 15, 16]
Housing buildings have a mashrabiya, an element of traditional Arab architecture used for limiting the income of radiation heat, through a wooden lattice whose openings allowed to cool the environment by evaporation of water placed in jars near them, currently in Masdar they are reinterpreted by wavy concrete mashrabiyas, which protect the windows and walls of radiation, avoiding thermal gains, while providing natural illumination and allows the passage of air refreshing the environments.
The design itself has elements of traditional Arab architecture, both the use of thermal masses and natural ventilation elements such as the atrium or windows oriented and placed in specific areas that, due to their location, provide a reduction in heat gains by radiation but that allow the passage of light that are now enhanced by the use of glass with low-emissivity technologies and with adjustable and

manipulable openings that also allow adequate ventilation regulated by the user, this without neglecting the need for a watertight environment that does not allow infiltration of humidity and heat from the outside environment [13, 16, 17, 18]

Another element rescued from the vernacular construction of the East are the wind chimneys, also called Bâdgir or ventilation towers, which have served as natural air conditioning for hundreds of years in Arab cities. These wind receivers have an orientation that allows the use of coldest breezes, allowing the entry of cold air and expelling the hot air, permitting to cool public areas such as squares and private areas like homes, even in some cases air can be directed through water tanks for better cooling.

In Masdar, the tower built in steel reaches a height of 45m, which, added to the blinds that direct the wind downwards and a cooling system that consists of moistening the air that descends towards the city with desalinated seawater, has managed to capture cold winds that direct towards the public square, achieving with this mixture of technology and tradition to provide a comfortable environment in the middle of the desert. [13, 16, 19]

3.3. Tianjin. Modern technological development

Located near the mega city of Shanghai in China, it is designed to reduce the population burden of this mega city, considering that China in addition of being a developing world superpower, that is why the levels of pollution and production of Co2 are among the highest in the world due to their levels of industrialization. It is also the most populated country in the world and suffers a constant migration to cities, so Tianjin is part of a series of sustainable cities with clear guidelines determined by 26 indicators (KPI) related to water, air, energy, transportation and waste, among its main aspects, and somehow are being "tested" in Tianjin, aimed at achieving a sustainable city that allows it to manage the particularities of its territory, such as the large amount of urban population it has and the fact that it continues to grow rapidly.[18, 20, 21]

Developed in partnership with Singapore, it has proposed sustainable strategies in order not only to provide comfort to its inhabitants through energy efficiency strategies using renewable energies such as solar, wind and geothermal energy, but seeks to regenerate its initially polluted natural environment, as it is built on an old landfill, surrounded by highly polluted water. To achieve this, the city has a vegetation corridor that crosses it, but also combines its 7 sectors (Lifescape, Eco-Valley, Solarscape, Urbanscape, Windscape, Earthscape and Eco-Corridors) so that vegetation is within reach for all of them, in addition the views that the city presents, are managed in a way that takes advantage of the existing or designed topography, it also makes efficient use of vertical developments so that they take advantage of strategies of the sectors of the city, offering users different forms to inhabit from an urban environment of high density and height (Urbanscape), to a relaxed and natural environment similar to the life of a town (Windscape), passing through an intermediate space providing its inhabitants with a natural and relaxing environment typical of the Chinese culture with large public and green spaces (Earthscape) [20, 21, 22]

The city is a high-density development; however, it is framed in a compact way to provide its inhabitants (350,000 permanent residents and 60,000 temporary residents in 34.2 km2), the possibility of close access to each of the spaces itself, thus reducing the need for a car, which is also enhanced by the development of pedestrian cycle paths, and an electric car park, as well as an excellent public transport system that is also electric [18, 20, 21, 22]

As mentioned in previous paragraphs, sustainability is not well defined and encompasses many characteristics within its concept, which intersects with smart cities that achieve sustainability through strategies with strong support from technology, as in this case, without forgetting the fundamental passive strategies to achieve energy efficiency, such as orientation. The city is oriented so that the openings such as windows face South in order to obtain more hours of sunshine and with this greater natural lighting reduces energy use, while the north side of the buildings has small windows that allow the entry of lighting, but limit thermal losses, maintaining the air conditioning of the interior environment, which they achieve with the support of materials such as glass, thermal insulators, or intelligent self-regulation systems, which have the most advanced technologies to achieve greater energy efficiency [20, 22]

In this city we can find technological advances such as the power generators (solar, wind and geothermal), already well known in sustainable environments, and which in the case of Tianjin cover at least 20% of the city's demand, but we also find new strategies, which, although they are in the

development phase, are surprising for their potential and ingenuity, as is the case of systems that take advantage of the forces of nature, generating small tornadoes for energy production. None of the needs of the city is forgotten, in this way the cleaning of the water is carried out by means of innovative engineering methods that, using textile materials, allow cleaning the water and together with other non-traditional sources, the city is overwhelmed by 50 % of the water supply necessary for its operation [20, 21, 22]
The city of Tianjin mixes passive strategies such as orientation, lighting, ventilation or insulation already mentioned, which has even led it to obtain, through the Eco City complex, the Passive House certification, as the first complex of passive housing in height of the world, however the strategies of traditional Chinese architecture are not a relevant element in the design of the city, except for orientation, since its characteristics and strategies respond to the requirements of its environment, which in this case are marked by the needs of its growing population, which drives city builders to use, promote and create new systems and strategies [20, 21]

4. Discussion

While Masdar faces an extremely hot climate environment, Tianjin faces the challenge of an excessively polluted environment with more heterogeneous population characteristics and greater prospects for constant growth. In this way, the strategies and elements to be used in these cities will differ according to the characteristics of their climatic, sociocultural and population environment. The city of Tianjin has been selected precisely as a contrast to the city of Masdar since, even though both use a large number of new technologies to achieve sustainability, in this case focused on strategies that provide comfort, Tianjin promoted the development of new technologies in addition to using existing ones, while Masdar combined these technologies, enhancing them with traditional strategies, achieving in both cases, creating efficient cities.

5. Conclusions

Both, Tianjin and Masdar are cities created from a political initiative, in countries with low levels of energy efficiency in their cities, and with high levels of consumption of fossil fuels that have recognized the new determinants of city planning and that have launched the trials of a sustainable city with these model projects.
Tianjin is a city that will support a high population density that will grow constantly and it is for this reason the strategies of traditional architecture are not a relevant element in the design of the city, except for the orientation, however, its characteristics and strategies respond to the characteristics of its environment, which in this case is marked by the needs of its population, to which it has responded with constant innovation and development of strategies based primarily on China's technological potential [18, 20, 21, 22]
One of Masdar most notable architectural "innovations" is in its design, a mixture of tradition and modernity, in which elements of the vernacular construction of the East and current technology are combined, reinterpreting knowledge of times when it was not thought of energy expenditure, but nevertheless, it was minimal since its architectural elements, as well as its materials, were a response to the location, climate and needs of each region, and were abandoned to imitate models of cities that were not for technology today, they would never have been built in places with such extreme climates [15, 16, 17, 19]
There is still a long time to be able to put the model label on these cities, and know exactly how efficient and sustainable they are, since we will only know once they are finished and working at full capacity, with all the pros and cons of the technologies implemented in the city, however, the innovations presented in this article are totally positive and not only applicable to these cities. Power generation systems, the planned public space environment based on the needs and characteristics of the population and the rescue of elements of traditional construction, are strategies that anywhere on the planet would help to create sustainable spaces, since as mentioned in one of the sections of the article, a sustainable city bases its design on the needs of each region and not only on aesthetics, which should not necessarily be sacrificed, but which must be combined creating unique environments in which as in antiquity recognizes the identity of the peoples [18, 20, 21, 22]

IOP Conf. Series: Earth and Environmental Science 1006 (2022) 012013 doi:10.1088/1755-1315/1006/1/012013

6. Reference

[1] Onu hábitat *2018 La gestión de residuos sólidos municipales* Objetivos del desarrollo sostenible ODS [Online] Avaible: https://onuhabitat.org.mx/index.php/dia-mundial-del-habitat-2018

[2] European commission 2019 Guidance on a strategic framework for further supporting the deployment of EU-level green and blue infrastructure (Brussels, Belgium SWD) p 1

[3] World Economic Forum 2020 *The Global Risks Report 2020* Marsh and McLennan and Zurich Insurance Group (Zurich, Alemania) 15

[4] Unated nations development program 2021 *Objetive 1:Sustainaable cities and communities* [Online] Avaible: https://www1.undp.org/content/undp/es/home/sustainable-development-goals/goal-11-sustainable-cities-and-communities.html

[5] West S, van Kerkhoff L, and Wagenaar H, 2019 Beyond linking knowledge and action: towards a practice-based approach to transdisciplinary sustainability interventions, *Policy Studies* 40(5) 534–55

[6] Hemström K, Simon D, Palmer H, Perry B, and Polk M. 2021 Transdisciplinary Knowledge Co-production: A Guide for Sustainable Cities (Rugby UK) *Practical Action Publishing* c 2 pp 51-55

[7] Zareba A, Krzeminska A, Lanch J 2017 Energy sustainable cities. From eco villages, eco districts towards zero carbon. *E3S Web of Conferences* **22**(1)

[8] United Nations, Department of Economic and Social Affairs, Population Division (2019). *World Population Prospects 2019*, Volume II: Demographic Profiles (ST/ESA/SER.A/427)

[9] Badach J, Dymnicka M, Baranowski A 2020 Urban vegetation in air quality management: a review and policy framework. *Sustainability*, **12**(3) L1258

[10] To K, Fernández J (2014) Alternative Urban Technology for Future Low-Carbon Cities: A Demonstration Project Review and Discussion *Springer* **10**

[11] Šlogar H, Bandov G, Čakanić T, 2020 *A sustainable city in the context of urban development*, International Scientific Conference ERAZ 253-257

[12] Amin Al-Habaibeh 2019 *Architectural lessons for the future, via the pas*t Unated Nations Educational, Scientific and Cultural Organization UNESCO [Online] Avaible: https://es.unesco.org/courier/2019-4/arquitectura-tradicional-fuente-inspiracion-ciudades-del-futuro /

[13] Ibrahim I, 2016 Livable Eco-Architecture Masdar City, Arabian Sustainable City *SciencesDirect* **216** 46-55.

[14] Foster and partners 2014 Masdar [Online] Avaible: https://www.fosterandpartners.com/projects/type/?projecttype=urban-design

[15] Masdar a mudabala Company 2021 *Sustainability Masdar City is one of the world´s most sustainable urban communities* [Online] Avaible: https://masdar.ae/en/masdar-city/the-city/sustainability

[16] Wheeler A, Rajan A, 2016 *The Educational Value of Masdar as an Exemplar Sustainable City: A critical study* International Conference on Passive and Low Energy Architecture (Los Angeles USA) **32**

[17] Picchio F, Bercigli M, De Marco R, 2019 Digital Scenarios and Virtual Environments for the Representation of Middle Eastern Architecture *Springer International Publishing* 44

[18] Wang X, Mell I, 2019 International Development Planning Review/ Evaluating the challenges of eco-city development in China: a comparison of Tianjin and Dongtan eco-cities *International Development Planning Review* **41**(2) 215–242.

[19] Madakam S, Ramaswamy R 2016 Sustainable Smart City: Masdar (UAE) (A City Ecologically Balanced). *Indian Journal* of *Science and Technology* **9**

[20] Li Y, Commenges H, Bordignon F, Bonhomme C, Deroubaix J, 2018 The Tianjin Eco-City model in the academic literature on urban sustainability *Journal of Cleaner Production*

[21] Joss S, Molella A, 2013 The Eco-City as Urban Technology: Perspectives on Caofeidian International Eco-City (China) *Journal of Urban Technology* 20 115–137

[22] Yang W, Deng W 2013 Building related KPIs in Sino-Singapore Tianjin Eco-City *APCBEE Procedia* **5** 112 – 115

AUTHOR INDEX